The Patient Flower

The Hidden Lives of North America's Native Lady's Slipper Orchids

Joel Zyla

Pine Duff Press

Published by Pine Duff Press.
First edition, 2026.
Paperback ISBN: 979-8-9965478-1-4
Hardcover ISBN: 979-8-9965478-0-7
Ebook ISBN: 979-8-9965478-2-1
This book is offered in the spirit of wonder and conservation. It is not a guide to collecting, growing, handling, or consuming wild plants, and nothing in it should be read as such. Native lady's slipper orchids are protected in many places and easily harmed; the author asks readers only to look, and to leave them where they grow.
Illustration and photo credits appear at the back of this book. Cover design by Pine Duff Press; cover art generated with artificial intelligence and verified by the author.
Printed in the United States of America.

For my grandfather, Merrill Carnes,
who took me outside first, and taught me to look.

Contents

Author's Note

A word before we begin, about how this book is made and where its limits are.

This is a book of wonder before it is anything else. It is not a manual for growing lady's slippers, and it is not a guide to finding and digging them; if you came for the latter, I will spend much of the book trying to talk you out of it. What I want is for you to see these flowers more clearly, because seeing them clearly and wanting to keep them turn out to be the same act.

The first thing that needs saying is about names. Every one of these orchids carried names, and lived in relationship with people, long before a botanist pinned a Latin binomial to a dried specimen. Across much of the northern range of the pink lady's slipper it is the moccasin flower, a name carried in Anishinaabemowin in forms built from *mak-izin*, moccasin or shoe; other Nations have their own names and their own long relationships with these plants. Where the published record gives me such a name, I use it, and I try always to do three things: to name the specific Nation rather than reaching for a vague "Native American," to name the published source or knowledge-holder the name comes from, and then to stop. I am a settler and a learner here. Not a translator of this knowledge, and not an authority on it. There are stories held with some of these flowers, and meanings, and uses, that are not mine to retell, and I have not retold them. I have tried only to say that they exist, to point toward their keepers, and to leave the door respectfully closed.

I did not invent this way of working. I learned it from Indigenous writers and scholars, Robin Wall Kimmerer first among them, and from the people and frameworks now doing the patient work of Indigenous cultural and data governance: Local Contexts and its Traditional Knowledge Labels, the CARE Principles, the OCAP principles in Canada. The short version of all of it is the oldest courtesy there is. Some knowledge belongs to a people, and the right way to treat it is to ask, to credit, and to defer.

More than once in these pages I note that the published record holds no name or no use for a given orchid among a given people. I have tried to frame those silences as what they are, gaps in a settler archive, and never as proof that the knowledge does not exist. A silence in our records is a fact about our records. It is nothing more than that.

I have checked the Indigenous-related material in this book against primary published sources as carefully as I know how, but careful citation is not the same as the blessing of the people whose knowledge this is, and I claim only the former. If I have gotten something wrong (a name, an attribution, a framing, an emphasis), I want to know, and I will fix it in the next printing. That offer is genuine and it has no expiration. You can write to me at joelzyla@gmail.com. You might think of this book as one settler's attempt to pay attention well, offered in the hope of being corrected where it needs correcting.

A word about the science. I have not invented anything in these pages; every claim about range, habitat, bloom time, pollinator, fungus, or conservation status traces to the published record, and where that record is uncertain or actively disputed I have tried to say so rather than smooth it over. Part of the joy of these orchids is how much about them remains genuinely unknown — pollinators no one has yet watched, fungal partners no one has yet named. I have tried to mark the live edge of what we know rather than pretend the edge is not there.

And a word about places. You will notice that I name regions and habitats but almost never a precise location. That is deliberate. For a rare orchid the single most dangerous fact is its exact address, and the internet is not the right keeper of it. If you go looking, go gently, and keep what you find the same way.

That is the whole of the housekeeping. The rest is the flowers.

Part One — Falling in Love

Before any of the science, there is the kneeling. This first part is about meeting these orchids the way most of us do, at close range and a little by accident, and finding that the flower which stopped you is stranger, and more cunning, than it looks.

The Flower Running a Con

Kneel down.

I mean it. Actually get down on your knees, because that is the only way to meet this flower on its own terms, and because the thing I am about to show you cannot be seen from standing height. It is late May in an oak-heath woods, the kind with white pine and pitch pine threaded among the oaks and last autumn's leaves still matted underfoot, and the light is coming through the half-opened oak canopy in that thin green way it does for only a few weeks a year. Push aside the leaf litter at the base of a pine. Take your time. And there, rising on a single bare stalk from between two pleated leaves, is a pouch of impossible pink, veined and swollen and nodding slightly in air too still to feel.

This is a pink lady's slipper (*Cypripedium acaule*). Some people call it the moccasin flower. I still go quiet every time I find one, the way you go quiet in front of something that has been at this far longer than we have.

A single pink lady's slipper (Cypripedium acaule) *on its leafless scape at the foot of a pine, in the thin green light of late May.*

Here is the first thing to understand, and it is the thing this whole book turns on. What you are looking at is not a flower trying to be beautiful for you. It is not even, really, trying to be beautiful. It is running a con. A very old, very good, very specific con — and you, kneeling there charmed, are not even the mark. You are just a passerby who happened to catch the show. The mark is a bee, and she has no idea what is about to happen to her.

What you are actually looking at

Let me name the parts, because the names matter and I will use them for the rest of the book.

That swollen pink pouch (the part that gives the plant every one of its common names, slipper and moccasin alike) is a single, wildly modified petal. Botanists call it the **labellum**, and I will mostly just call it the pouch or the slipper, because that is what it looks like and that is what people have always called it. Most flowers wear their petals flat and open, like a hand held out. This orchid took one petal and inflated it into a hollow chamber, a balloon of tissue with an opening in front, and arranged everything else around that one astonishing structure: the other petals, the sepals, the reproductive machinery. The whole flower is built to serve the pouch.

The pouch sits at the top of a single bare stalk. In C. *acaule* this stalk carries no leaves at all. It rises naked from the ground, and the leaves stay below, two of them, pressed close at the base. That leafless stalk has a name too: a **scape**. (The plant's scientific name, *acaule,* means "stemless," and it is one of the few orchid names that is honest about what it describes — though more on that in a later chapter.) The leaves are pleated like a folding fan, ribbed with parallel veins, soft and faintly hairy to the touch. Two leaves, opposite each other, hugging the ground. Hold that detail; it will turn out to be the single most reliable way to know this orchid from every one of its cousins.

The two opposite basal leaves, pleated and parallel-veined, with the leafless scape rising between them, here topped by a dry seed capsule: the surest way to know the pink lady's slipper.

Now look at the front of the pouch.

Every other lady's slipper in North America, every one of the eleven other native species in this book, opens its pouch through a round hole near the top — like the mouth of a little jug. Not this one. *C. acaule* opens down the front, through a long vertical slit that runs the length of the pouch, and the two edges of that slit are rolled inward on themselves, curled back toward the inside of the chamber. This vertical fissure with its inrolled lips is found in no other *Cypripedium* on Earth. It is the flower's signature. And it is also the trap.

The mark walks in

A bee comes. Specifically a bumblebee, and not just any bumblebee but, across most of this orchid's range, a queen of the half-black bumblebee, *Bombus vagans*, or in the colder northern parts of the range her amber-banded relative, the northern bumblebee *Bombus borealis*. I want to be precise about this, because for a long time the books got it wrong and blamed the wrong insect; the small solitary bees that pollinate the yellow lady's slippers do not do the work here. This pouch is built for something bigger. It is built for a queen.

She is drawn in by the color first, that hot pink visible against the brown forest floor from a surprising distance, and then by a faint sweet scent. She lands. She finds the slit, pushes at the inrolled edges, and forces her way through, because she is big and strong, a queen newly out of hibernation with a colony to found. The rolled lips give way and let her in. A smaller, weaker bee could not have managed it. That resistance is the first thing to notice: the slit is a filter, and it is sized for muscle. This flower is, in effect, auditioning its pollinators for strength, and no other lady's slipper on Earth opens this way.

And then they fold in behind her.

This is the part that should change how you see the flower. The inrolled margins of that fissure are not decorative. They flex inward easily as the bee pushes in, and then they relax back across the gap, so that the easy way she came is suddenly the hard way out. She is inside a smooth, translucent, dimly glowing chamber, and every cue the flower offers now points her somewhere other than back through that slit. Here is the heart of the con: there is nothing in there for her. No nectar. No reward of any kind. She has been drawn into a one-way room on a promise the flower never intended to keep.

This is what biologists call **deceptive pollination**: the flower advertises a meal it does not serve. Orchids are among the great cheats of the plant world, and a great many of them lure their pollinators exactly this way, offering color and scent and shape, every signal of a good flower worth visiting, then delivering nothing. The lady's slipper is one of the most accomplished practitioners. The bee works for free, and she does not even know she is working.

The only way out

So the queen is trapped in a chamber the size of a thimble, agitated now, looking for an exit. The walls are slick and offer no foothold. But the back of the pouch is translucent, lit from behind like frosted glass, and near the top of that back wall are two small bright windows where the light comes through strongest. She does what any animal in a dark room does. She climbs toward the light.

And the flower has built the only path out exactly where it needs her to go.

To escape, she has to squeeze up and through one of two narrow openings at the rear of the pouch, tucked beside the column where the flower keeps its reproductive organs. Forcing back out through the inrolled slit would take real work, and the lit rear of the chamber is the path of least resistance, so she takes it: forward, up, and out the back, through a corridor so tight she has to scrape her way along it. And the order in which she meets things in that corridor is not an accident. She passes the **stigma** first, the sticky surface that receives pollen, and only after that does she pass the **anther**, where the flower's own pollen waits to be glued onto her back.

Sit with that sequence for a second, because it is a small masterpiece. If she is carrying pollen from another lady's slipper she visited earlier, she leaves it on the stigma on her way past, pollinating this flower, *before* she picks up this flower's own pollen. The flower has arranged its anatomy so that incoming pollen is delivered before outgoing pollen is loaded. It is almost impossible for the plant to fertilize itself. Every grain that does its job has to come from somebody else's flower, carried on the back of a bee who was conned into the trip. The trap doesn't just use the bee. It uses her to force the orchid to outbreed.

She drags free at last, dusted with pollen, no richer than she came. And if the con works, she flies off and finds another pink pouch and gets fooled all over again.

A flower running a con

That is a flower running a con. And it is a very old one. The genus *Cypripedium* has been trapping bees in pouches for something like thirty million years, which means these chambers were luring insects to work for nothing long before there was anything resembling us to kneel down and admire the trick.

I should be honest about how well the con actually pays, because the numbers are part of the wonder. In two Virginia populations that were studied closely, researchers found that only about five percent of flowers in one population, and thirteen percent in the other, ever set fruit on their own — while flowers they pollinated by hand set fruit at essentially a hundred percent. The flower is fully capable of making seed. It just rarely gets the visit. A bee, once trapped and cheated, learns fast, and is in no hurry to repeat it. So most lady's slipper flowers, in most years, bloom gorgeously and are never pollinated at all. They are gambling on the rare naïve queen who hasn't learned better yet.

But here is the twist the same study found: when a bee *does* take the bait and carry pollen onward, she is a spectacularly good vector. Somewhere between a third and a half of the pollen she removed went on to fertilize other plants. The visits are rare and the visits are precious. The orchid has made a bet that a few well-aimed deceptions are worth more than a thousand honest little payments of nectar it would have to manufacture and give away. Across thirty million years, the bet has held.

Beauty was never the point — and that is the point

I told you that you, kneeling there, are not the mark. You aren't. The pink was never for your eyes; it was tuned, over deep time, to the visual system of a hungry queen bumblebee in a spring forest, and the fact that it moves us too is, biologically speaking, a coincidence.

But it is the most useful coincidence I know.

Because once you understand that the beauty is a tool, a precise and evolved and ruthless tool for getting a specific insect to do a specific job for free, the flower somehow becomes *more* beautiful, not less. The pink stops being mere prettiness and becomes a kind of argument the plant is making with its environment, a strategy worn on the outside. You start to see the slit not as a flaw or an opening but as a machine. You start to see the whole nodding pouch as what it is: not an ornament, but an instrument, played on bees.

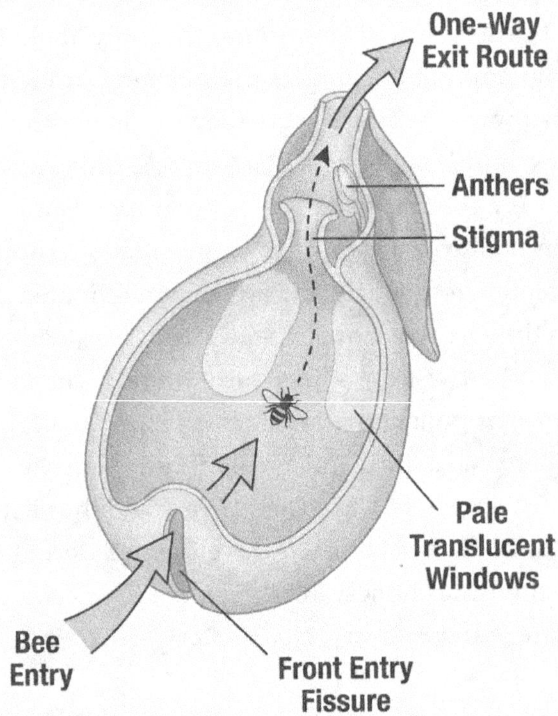

How the trap works. A bee forces in through the slit down the front, climbs toward the lit windows at the rear, and escapes only past the stigma and then the anther.

And then the questions start, the way they started for me, kneeling in the woods my grandfather first walked me into as a boy, and the way I hope they start for you. If the flower so rarely sets seed, where do new lady's slippers come from? When a seed does form, what happens to it? A single pouch, successfully fooled, can spill a haze of dust-fine seeds, each one barely more than a speck. How does a dust mote with almost no food of its own become a plant that takes six to ten years to flower for the first time and can then go on living for decades? What is going on under the ground, in the dark, all those years before the first leaf ever sees the light?

I will tell you, but not yet. For now I only want to plant the answer, because it is the secret at the center of this book. This orchid cannot grow alone. It never could. From its first hours as a seed in the soil, *C. acaule* is utterly dependent on a partner it cannot see and we usually forget: a particular fungus, threaded through the soil and into the seed itself, feeding the orchid in the years before the orchid can feed itself. We tend to call that kind of dependence fragility. By the time we reach the chapter where the fungus comes onstage, I am going to try to convince you it might be better to call it trust. The con the flower runs on the bee, it turns out, sits on top of a partnership the flower keeps, faithfully, with something underground — and that partnership is the more astonishing story of the two.

What we have called it

People have been kneeling in front of this flower for a very long time, and they have left us their words for it.

The naturalist Susan Fenimore Cooper, walking the woods near Cooperstown, New York, in the early summer of 1850, wrote in her journal that "we found also a little troop of moccasin plants in flower; frequently, the season has passed without our seeing one, but this afternoon we gathered no less than eighteen of the purple kind, the *Cyprepedium acaule* [sic] of botanists." I love that passage for the joy in it, the giddy bounty of eighteen at once, and I wince at it for the same reason, because the gathering of eighteen wild orchids is exactly the impulse this book exists to talk you out of. Cooper herself felt the loss coming. Earlier that same spring she had noted that the moccasin flower, once common right up to the edge of the village, had grown rare, "and it is considered a piece of good luck to find them." Within one woman's lifetime, in one corner of New York, abundance was already turning to luck. We will come back to that.

Henry David Thoreau, a few years later and a state away, kept finding it where I have always found it: on dry, acid, pine-shaded ground. "The lady's-slipper in pitch pine wood-side near J. Hosmer's Desert," he noted in his journal in late May of 1856, "probably about the 27th." That offhand line carries a real piece of natural history. Thoreau, without theorizing about it, had pinned this orchid to the poor, sandy, sour soils where it actually lives — the pine woods and oak heaths and bog margins, the ground other plants find too acid to bother with. The pink lady's slipper is a specialist in difficulty. It thrives in soil sour enough, with too little for most things to eat, that it has the place largely to itself. There is a whole chapter in that, too.

The moccasin flower has older names than the botanists' Latin and older claims than the naturalists' journals. In Anishinaabemowin, the language of the Anishinaabe peoples whose homelands stretch across much of this orchid's northern range, it carries names built from the word *makizin*, moccasin or shoe. The Great Lakes Indian Fish and Wildlife Commission records the name *makizinkewe*, "moccasin-making one," in its 2007 guide *Indinawemaaganidog / All My Relatives*; the root *makizin* is documented in the Ojibwe People's Dictionary at the University of Minnesota, and the fuller compound *makizini-waabigwan*, "moccasin flower," in contemporary Anishinaabe-language sources. That same root gives us the English "moccasin flower," a calque of the older Anishinaabe name, and one that persisted among most Americans until nineteenth-century European usage, "lady's slipper," displaced it. There is an Anishinaabe story tied to this flower, held by the people whose story it is, and carried still by Anishinaabe storytellers today. It is not mine to tell, and I am not going to retell it here. I want only to say that it exists, that it is older than this book and not made for it, and that I am writing as a settler and a learner who knows enough to leave that particular door respectfully closed. The flower has lived in language, and in relationship, far longer than it has lived in field guides. ## The whole game

I have watched people walk straight past a colony of these orchids without seeing a single one. The pink hides better than you would think against the leaf litter, and a person moving at standing height, talking, thinking about lunch, simply does not register it. I have also watched a person who finally *did* see one, who knelt the way I told you to kneel, go completely still. That stillness is the whole game. You do not protect what you have never noticed, and there are fewer of these flowers along our roadsides and in our woodlots every year, in part because so few of us are looking down.

So this is partly a book about a flower, and partly a book about learning to look.

The promise I will make you now, and try to keep for the next fifteen chapters, is small and specific: by the time we are done, you will not be able to walk past one of these again. You will see the pink. You will look at the slit down the front of the pouch and think, *one-way valve.* You will wonder what fungus is feeding the plant from below and how many years it waited underground before it dared to bloom. And you will understand, standing there, that the most beautiful thing on the forest floor is also one of the most cunning, and that the two facts are not in tension at all. They are the same fact.

For now, though, just stay down here a moment longer. Look at the pouch nodding on its bare scape, the two pleated leaves cupped at the base, the long inrolled slit catching the green light. Somewhere a queen bumblebee is doing her rounds, and the flower is waiting, patiently, with the deep patience of a lineage that has been at this for a very long time, for exactly the right dupe to come along.

Then stand up, brush the leaf litter from your knees, and leave it where it stands.

What Is a Lady's Slipper, Really?

Walk into any garden center in May, anywhere on the continent, and somewhere near the front, between the hydrangeas and the impulse-buy succulents, you may find a pot holding what the tag calls a "lady slipper." The flower is unmistakable. A pouch, swollen and waxy, the color of a bruise or a sunset, slung beneath a flare of glossy petals. It looks like the thing this whole book is about. You could carry it to the register, set it on a windowsill, and tell your friends you are growing a lady's slipper.

You would be growing an orchid. You would not be growing one of ours.

I want to start here, at the register, because it is the cleanest way I know to draw the line this chapter is about. The plant in that pot is almost certainly a *Paphiopedilum,* an Asian slipper orchid, or more likely a nursery hybrid bred from several Asian species and selected over decades for a pouch that holds its color under fluorescent light and a constitution that forgives a forgetful owner. It is a marvel. It is also a houseplant, the way a golden retriever is a wolf: same deep family, completely different life. The flowers this book follows do not sit in pots near the register. They grow in fens and pine woods and prairie remnants and serpentine seeps, they cannot be bought cheaply because they cannot be made cheaply, and most of them would be dead within a season if you dug one up and brought it home. To understand why, you first have to understand what a lady's slipper actually *is*: what makes it an orchid, what makes that pouch a pouch, and which "lady's slippers" are family and which are strangers wearing the name.

So let's take the flower apart. Gently, and without picking it.

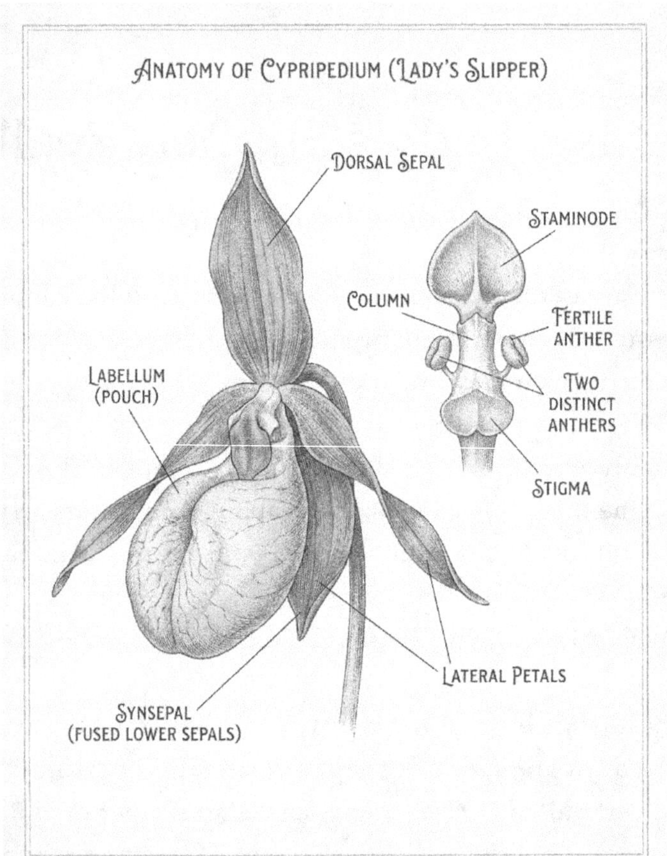

The parts of a lady's slipper. An inflated labellum, a hooding dorsal sepal, two lower sepals fused into a single synsepal, and a column carrying two fertile anthers: the build that marks the slipper orchids.

What makes it an orchid

Start with a thing you already know how to do without thinking. Look at your own face in a mirror. A line drawn straight down the middle, between your eyes, down the bridge of your nose, divides you into two halves that more or less match. Left eye, right eye. Left ear, right ear. This is **bilateral symmetry**, two mirrored sides, and you share it with nearly every animal you've ever met. Most flowers do not work this way. A buttercup, a wild rose, an apple blossom: spin any of them like a pinwheel and they look the same from any angle, petals arranged around a center like spokes on a wheel. Botanists call that radial symmetry, and it is the floral default.

Orchids broke the default. An orchid flower has a top and a bottom, a left and a right, a single plane down which it folds into matching halves: a face, not a wheel. This is the first and oldest tell. Once you have it in your eye you cannot lose it. The orchid on the windowsill, the lady's slipper in the fen, the little green twayblade you'd otherwise step on, the five-dollar grocery-store *Phalaenopsis,* every one of them is built like a face, with one petal singled out and dressed up to be the lower lip of that face. That singled-out petal is where the whole story lives. Hold that thought; we are coming back to it.

Now the part that is genuinely strange, the feature that more than any other says *orchid* and nothing else. In most flowers, the male organs and the female organs are separate neighbors. There are stamens, slender filaments tipped with pollen, standing in a ring, and somewhere in the middle a pistil with its sticky landing pad and its ovary below. Boy parts here, girl parts there, sharing the same blossom but distinct structures you could point to one at a time. Orchids fused them. In an orchid, the male and female machinery are welded into a single organ, one structure doing both jobs, and that fused organ has a name worth learning: the **column**. It is the defining hardware of the orchid family. No other group of plants makes one. When you see a column, you are looking at an orchid, full stop.

And then there are the seeds, which belong in any honest definition. Most plants pack a lunch for their offspring. A bean, an acorn, a sunflower seed, each is a baby plant wrapped around a generous store of food, enough fuel to push a root down and a shoot up and run the first solar panels before the seed needs the outside world for anything.

Orchids send their children out with the pockets empty. An orchid seed is dust. Not dust as a figure of speech. Dust as a fact, a speck so small and so light that a single ripe seedpod releases them in their hundreds of thousands and the wind takes them like smoke. They carry almost no food at all.

That empty lunchbox has a consequence, and it is the last item on the checklist: an orchid seed germinates only after a soil fungus moves in and feeds it, for months or years, before the first leaf. The dust seed lands, waits in the dark, and a particular fungus threaded through the soil finds it and grows inside it, passing sugar and minerals across a living seam until the orchid can finally make a leaf of its own. The relationship has a name, **mycorrhiza**, and it is the real engine under everything else a lady's slipper does. There is a whole chapter coming on it, the one where the fungus comes onstage. Not yet. For now I will just set it on the table and move on.

So: a face, not a wheel. A column where most flowers keep their sexes apart. Seeds like smoke, sent out broke. And a hidden partner underground without which none of it happens. Stack those four together and you have the orchid family, the largest family of flowering plants on Earth, tens of thousands of species in nearly every habitat that isn't open ice or open ocean. Our lady's slippers are a tiny, ancient corner of that enormous house. To find their corner, look at the lip.

The slipper is a petal

Picture two labellums side by side. On the windowsill *Phalaenopsis,* the lower lip is a small frilled landing platform, a flat little apron a moth could perch on. On a *Cypripedium,* it is a balloon you could lose a bumblebee inside. Same petal, same ancestry, wildly different ambition, and the lady's slipper's version is the whole reason anyone ever gave this plant a name about footwear.

That pouch is one petal. A single petal, and nothing more exotic than that. Other flowers wear their petals flat, splayed open like a hand. The lady's slipper took one of its three petals and blew it up like a glassblower working a bubble, inflated it into a hollow chamber with an opening you could slip a fingertip into, and there is your slipper. The botanical name for that transformed lip is the **labellum**, Latin for "little lip," and it is the same modified lower petal every orchid carries, the singled-out one I told you to hold in your eye.

Understand what you are actually seeing. That pouch is not a decoration and not an accident of beauty. It is a *room*, a chamber a pollinator enters and cannot easily leave the way it came, with the flower's reproductive equipment positioned at the exits so the visitor has to brush against it on the way out. The shape is the strategy. We took the long route through that strategy in the first chapter; here just hold the pouch in your mind as architecture, not ornament, and the rest of the flower's parts will make sense as the walls and signage around that one decisive room.

Because the labellum does not float there alone. Look above it and behind it. A lady's slipper flower has six parts in the old orchid arithmetic, three sepals and three petals, with the genus doing something specific and recognizable with each.

Above the pouch, usually held up like a hood or an awning, is a single broad **sepal**, the **dorsal sepal**, the one most flowers would keep at the back of the bud. In the lady's slipper it stands tall over the opening, and in many of our species it is the showiest flag the plant flies, a banner of color advertising over the pouch's doorway. Below and behind, where you'd expect to find the two remaining sepals, the lady's slippers fused them into a single structure that hangs beneath the pouch like a second, narrower hood pointed downward. Botanists call that fused pair the **synsepal**, and once you know to look for it, that downward-pointing "two sepals acting as one" beneath the slipper becomes another reliable way to know you're holding a *Cypripedium* and not something pretending. Then the two remaining true petals, the side ones: in most of our lady's slippers these stream out to either side, sometimes flat and pointed, sometimes spiraling like ribbon, framing the pouch left and right. Dorsal sepal up top, synsepal below, two petals out to the sides, and the third petal, the labellum, inflated into the pouch in the middle. That is the whole face.

Which leaves the column, hidden in the doorway of the pouch where the real business happens. I told you the orchid column fuses male and female into one organ. The slipper orchids do it with a peculiar twist that sets the whole subfamily apart from every other orchid on the planet. Most orchids carry a single anther and bind their pollen into hard waxy packets called pollinia. The slipper orchids keep **two fertile anthers**, one on each side of the column, and their pollen stays mealy and granular rather than packed into pellets. It is an old arrangement, a holdover from deeper in orchid history, and it is so consistent that botanists use it to draw the line around the whole slipper group. And between those two anthers, capping the column where a third anther would sit in some ancestor, the lady's slipper grows a shield: a sterile, often beautifully colored plate called the **staminode**, literally a stamen that gave up being a stamen.

The staminode is the part I find hardest to walk past. It makes no pollen. It does something arguably cleverer. It sits at the back of the pouch like a sign hung over a door, patterned and sometimes glistening, and it appears to help steer the trapped pollinator toward the exits, toward the anthers, toward the work the flower needs done. Some of our species carry a staminode dabbed with spots, some a clean bright yellow, some a strange pearled texture that catches the light, and the differences are sharp enough that the staminode alone can tell two look-alike lady's slippers apart. A stamen that quit reproducing to become a piece of signage. The slipper is full of these reassignments: a petal that became a room, a stamen that became a shield, a goddess's name for a swollen lip. Almost nothing about it is doing the obvious thing.

A petal for a pouch. A dorsal sepal for an awning. Two sepals fused into a hood. A column carrying two anthers and a billboard where a third anther used to be. Learn those six parts on one flower and you can read every lady's slipper in this book, and tell it, in a glance, from the impostors.

The four that aren't ours

So let's name the impostors, fairly, and then close the door on them for good.

The slipper orchids are not a single genus. They are a whole subfamily, an old, distinctive branch of the orchid family, set apart by that pouch and those paired anthers. Within it sit five genera, five sibling lineages that all make a slipper. Four of them are tropical. One is ours. The four tropical ones are worth knowing by name, precisely so you can recognize them as *not* the subject of this book the next time one turns up on a windowsill, in a florist's cooler, or in a glossy orchid-show photograph captioned, maddeningly, "lady slipper."

You have already met one of them at the register. ***Paphiopedilum*** is the slipper orchid of commerce, the genus sold under the trade name "Venus slipper," native to southern China and across tropical and subtropical Asia. There are something like a hundred species of it in the wild and tens of thousands of registered hybrids on top of that, bred and crossed and re-crossed for the nursery trade. When a tag, an orchid society, or a gift-shop sign says "lady slipper" and there's a price on it, the overwhelming odds are that you are looking at a *Paphiopedilum* or one of those countless man-made crosses. They are gorgeous, they are growable, they have launched a thousand obsessions. They are also, every one of them, foreign to North America. None grows wild on this continent.

Long ribboning petals that hang from the flower like streamers, sometimes dramatically so: that is the calling card of ***Phragmipedium***, the slipper orchids of tropical America, ranging from southern Mexico down through tropical South America. They are the showpieces of a certain kind of collector. And here is the line that matters for us: not one *Phragmipedium* species is native to the United States or Canada. The whole genus stays south of the border. A spectacular slipper orchid, a real one, a wild one, and still not one of ours.

Then there are the two almost no reader will ever meet. ***Mexipedium*** is the rarest kind of thing in botany, a genus with a single member, *Mexipedium xerophyticum,* known from a tiny wild population in Oaxaca, in southern Mexico, one species clinging to one corner of one country. ***Selenipedium*** goes the other direction entirely. These are the giants of the family, slipper orchids of tropical Central and South America that abandon the modest stature of every other slipper and grow on canelike stems, the largest of them, *Selenipedium chica,* reaching something like four to five meters (thirteen to sixteen feet), the tallest slipper orchid in the world. Roughly six to ten species, rarely cultivated, rarely seen, in rainforests you will probably never stand in. Mexican and South American, scientifically fascinating, and not ours.

Four tropical genera. Four kinds of slipper that are not the subject of these pages. I am not waving them off because they don't matter. They matter enormously, and a couple of them are in worse conservation trouble than some of ours. I am waving them off because confusion about the name has done real harm to the plants this book *is* about. The *Paphiopedilum* at the garden center makes a fine houseplant, and people learn that the easy way, on a windowsill. The trouble is that they carry that ease over to the wild natives, and they dig. They assume that because one kind of lady's slipper transplants happily into a pot, the one in the woods will too. It will not. So I want the four tropical names sitting in your memory like a closed door: those are the slippers you can own. Now let's open the other door, the one this whole book lives behind.

Back to the genus that is ours

Cypripedium. Say it once, because it is the only Latin name you truly need to leave this chapter with. Every native lady's slipper in North America north of Mexico, all twelve of them, belongs to this one genus, and only to this one. *Cypripedium* is the temperate slipper, the cold-country slipper, the one that left the tropics behind. While its four tropical siblings stayed in the heat, this lineage spread across the colder top of the world: through the woods and wetlands of North America, across Europe, deep into Asia, a band of slipper orchids ringing the temperate Northern Hemisphere. Botanists have a word for that sweep of distribution, *Holarctic,* the great northern ring of the globe, and most of *Cypripedium*'s roughly fifty species ride it, with a single arm of the genus reaching south through the mountains of Mexico and Central America as far as Honduras. Our twelve are the North American chapter of that much larger, much older story.

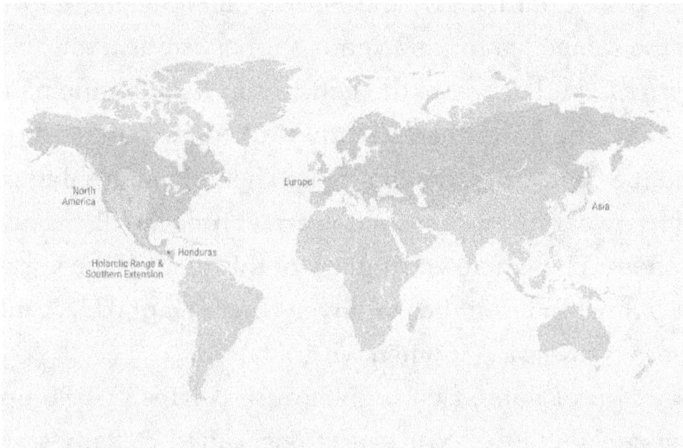

The genus Cypripedium *rings the cold top of the world, across North America, Europe, and Asia, with a single arm reaching south to Honduras. Botanists call that pattern Holarctic.*

And they are old. This is the fact I most want you to carry out of an orientation chapter, because it reframes everything that comes after. The genus *Cypripedium* is some thirty million years old; its lineage traces back to the Oligocene, an epoch with no grass prairies as we know them, no humans anywhere on the horizon, a world whose maps would be unrecognizable. The slipper orchids as a whole, that entire pouch-and-paired-anthers subfamily, split off from the rest of the orchid family roughly sixty-six million years ago, at the K–Pg boundary, the same geological moment that closed out the non-avian dinosaurs. Whatever else a lady's slipper is, it is the survivor of a lineage on a timescale that should make us quiet. The pouch you knelt to look at follows a design older than almost every living thing you can name.

There is a pleasure, too, in the name itself, because the people who named this plant heard the same thing in it that you and I do. *Cypripedium* comes from the Greek *Kypris,* an old name for Aphrodite, "the lady of Cyprus," the goddess the Greeks and Romans worshiped as the embodiment of beauty and desire, joined to *pedilon,* a word for a sandal or slipper. Venus's slipper. The little pouch was a goddess's shoe, set down at the edge of a wood, and the genus carries that meaning in its very roots. (Confusingly, the *Paphiopedilum* at the garden center borrows the same idea for its "Venus slipper" trade name, but it is the wild *Cypripedium* that wears the goddess in its scientific name.) Long before botany, the flower had already walked into language wearing a name about a beautiful woman's shoe. The folk names that came later all heard footwear too: moccasin flower, whippoorwill's shoe, two-lips, squirrel shoes. The whole vocabulary of this plant is the vocabulary of a slipper left behind. We did not invent the metaphor. We inherited it. The flower was wearing it when we got here.

One last bit of name-clearing, because it trips people up and the fix is a single sentence. Not everything called a "lady's slipper" is even an orchid. In some parts of the country people call jewelweed (*Impatiens capensis*) "wild lady's slipper," the orange-flowered plant of wet ditches and stream banks whose dangling, spurred blossom apparently struck somebody, once, as slipper-shaped, but it is no orchid at all. It is a fine plant. It is not a *Cypripedium;* it is not in this book except in this sentence. If someone points you to a "lady's slipper" and it is hanging in a bright orange cluster over a wet roadside in late summer, smile and keep walking. The flower you're after is rarer, slower, and stranger than that.

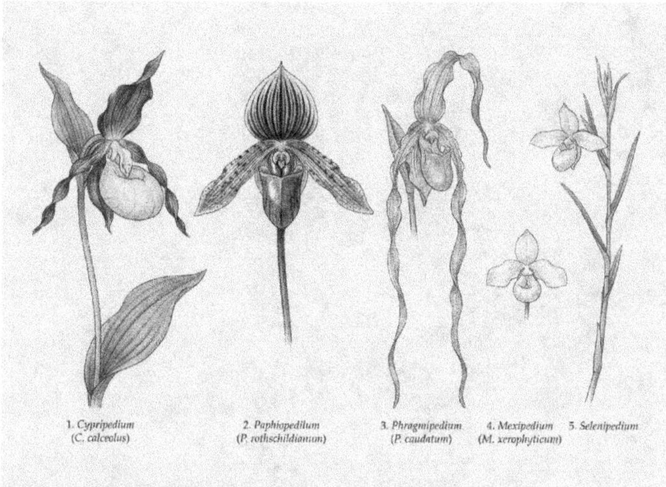

One bloom from each of the five slipper-orchid genera. Only Cypripedium, *at far left, grows wild in North America; the other four are tropical.*

So: *our* lady's slipper is a *Cypripedium,* not a *Paphiopedilum,* which is the difference between a thirty-million-year-old wild orchid of the cold north and the lovely tropical hybrid on the windowsill, a difference that decides, among other things, whether the plant survives being moved. There are twelve of ours, and the rest of this book walks you out to meet them.

The plant at the garden center: a potted Paphiopedilum *hybrid, the tropical impostor sold as a houseplant. It is not one of our wild natives.*

The next time you pass that pot at the front of the garden center, you'll know exactly what it is, and exactly what it isn't. But leave the register knowing that the flower this book is about is still out there in a fen somewhere, on a stem it took a decade to raise, holding a pouch that a goddess could have stepped out of, and waiting, as it has waited since before there were people to walk past it, for the right small visitor to climb inside.

Part Two — The Hidden Life

Almost everything that matters about a lady's slipper happens where you cannot watch it: in a dust seed with no food of its own, in a fungus threaded through its roots, in the slow underground years before a first green leaf. This part goes down into the dark, to the partnership the whole genus is built on, and to the patience it takes to become a flower at all.

The Dust That Dreams of Being a Flower

It is late summer now, months past the bloom, and the flower I knelt in front of in the first chapter is gone. What is left where the pouch hung is a dry, ribbed pod the length of your thumb, split along its seams and gone the brown of a paper bag. This is the capsule, the seed case, and almost no one ever notices it. The pink was the spectacle. This is the aftermath, easy to walk past, easy to mistake for a dead twig.

So hold out your hand, palm up, in good light. Tip the pod over your open palm and tap the side, gently, the way you'd tap salt from a shaker.

What comes out is not seeds in any sense you are used to. What comes out is dust.

A pale tan haze of it, so fine it seems less like a thousand small objects than like a single substance, a powder, a breath of smoke that happens to have settled on your skin. You cannot make out the individual grains without help. Each one is roughly a millimeter long and weighs a few millionths of a gram, which is to say it weighs almost nothing at all. A single pod holds them in their thousands: in the European yellow lady's slipper, *Cypripedium calceolus*, the Old World sister of our native yellows, careful counters have opened one capsule and tallied somewhere between six and seventeen thousand seeds inside it. Breathe out through your nose and you will lose some off your palm to the air. That is exactly how the plant intends to lose them. The wind is the whole delivery system, and dust is what the wind can carry.

Now look closer, because what is inside one of those specks is among the strangest facts in this book.

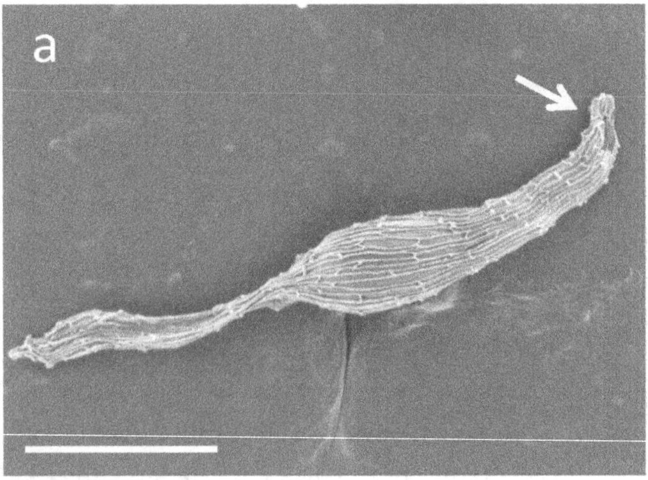

An orchid seed under the microscope: a loose papery coat around a minute embryo, carrying almost no stored food. (Shown here in Bletilla*, a representative orchid.)*

Dust with ambitions

I want to slow down here, because we move past this too quickly. We hear "tiny seeds" and we nod and keep reading, and we miss what tiny actually means.

Think about the seeds you know. A bean. An acorn. A sunflower seed, a grain of corn, the pit inside a peach. Every one of them is built on the same sensible plan: a baby plant, the embryo, wrapped around a packed lunch. That lunch has a name, **endosperm**, the starchy or oily food store a mother plant lays down so her offspring can get started. The acorn is mostly lunch. The bean you eat is almost entirely lunch, two thick seed-leaves swollen with enough starch to drive a root downward and a shoot upward and unfurl the first true leaves before the seedling has to find a single thing in the outside world. A normal seed is a baby with a sandwich and a thermos, sent off prepared.

An orchid seed packed no lunch.

Open one up, if you could, and here is what you would find. A seed coat, the **testa**, which is not the thin skin of a bean but a loose, transparent, papery shell, often inflated with a pocket of air that helps the whole thing drift. The botanists who have looked at it closely describe this coat as built in layers, a hardened outer wall and an inner armored skin, a kind of carapace, lignified and tough, and that toughness has a cost we will come to in a moment. And then, rattling around inside that papery shell, suspended in the empty space: the embryo itself. Not a folded seedling. Not even a recognizable plant. A blob of a few dozen undifferentiated cells, no root, no shoot, no leaf, nothing yet decided. Tucked among those cells, a scatter of oil droplets, a few microscopic beads of fat, and that is the entire pantry. No endosperm. No starchy lunch. No thermos. A handful of oil drops to a clump of cells that has not even decided which end is up, sealed inside an armored shell and thrown to the wind by the thousand.

The plant could have made a few dozen well-provisioned seeds, each one a fat acorn of an orchid with a real chance of getting started on its own. Instead it makes thousands of motes, and pours into each one almost nothing. It traded the lunchbox for sheer numbers. It is buying lottery tickets, by the thousand, and printing each one on a flake of ash.

Here is what that costs. A seed this empty cannot start itself. The oil drops are not enough fuel to crack that armored coat, push out a root, and build the first green leaf that would let the seedling feed itself on sunlight. There is no runway. A bean has the reserves to get airborne before it needs the world; a lady's slipper seed does not have the reserves to get off the ground at all. Left to itself on the forest floor, a *Cypripedium* seed does the only thing it can do, which is nothing. It lands, and it waits, and on its own it waits forever. It is a baby plant that physically cannot be born unaided.

So how does it get born?

The thing the seed is waiting for

It gets found.

This is the hinge the whole life turns on, and it is so improbable that when you first really understand it you may not quite believe me. A lady's slipper seed does not germinate on its own because it cannot. It germinates only if, lying there in the dark of the soil, it is discovered by a fungus: a particular kind of soil fungus threaded invisibly through the leaf litter and the mineral earth, going about its own life, and only if that fungus does two specific things for the helpless speck it has found.

First, the fungus opens it. That armored, lignified seed coat I described, the tough carapace that keeps the dry seed safe for years, is also a locked door, and the embryo inside has no key. The fungus does. It grows up against the seed and works its way in through the coat, breaching the shell from outside, cracking open a chamber that the embryo could never have opened for itself. The lock that protects the seed is the same lock that imprisons it, and the fungus is the thing that picks it.

Second, the fungus *feeds* it. Remember, the embryo has no lunch and no leaf, no way to make food and no stored food to live on. So the fungus supplies it. Carbon, sugars, the raw fuel of being alive, passed from the fungal threads across a living seam directly into the orchid embryo, in the dark, underground. The orchid does not photosynthesize. It cannot; it has no green tissue yet, and it will have none for a long time. For now it eats the fungus's groceries. A plant that lives by consuming a fungus instead of by catching the sun has a name, and the name is worth learning because it describes the entire childhood of every lady's slipper that has ever lived: **mycoheterotrophy**, fungus-feeding. *Myco* for fungus, *heterotroph* for an organism that gets its carbon by taking it from something else rather than making its own. The orchid begins life as a fungus-eater.

I am going to keep the fungus itself slightly offstage for now, because it deserves a chapter of its own and it is coming next. For the record: it belongs to a group of soil fungi the books call the Tulasnellaceae, an unglamorous, ground-dwelling lineage you have walked over ten thousand times without a thought. Who that partner is, how the relationship actually works, what each side gives and gets, whether to call it a deal or a theft or something with no good word for it at all: that is the next chapter, and I will not spend it here. What matters in *this* chapter is narrower and, I think, even stranger: that the orchid cannot get off the starting line without it. Not "does better with it." Cannot.

From its first hours, before it is anything we would call a plant, the lady's slipper is kept alive by a fungus. Full stop. Whether that is fragility or something better is the next chapter's argument, not this one's. For now the fact stands bare and a little alarming: the most helpless seed in the forest survives only by depending, completely and immediately, on a stranger underground.

And most of them are never found.

We should be honest about that, because it is the cost of the whole strategy. Thousands of dust seeds blow off a single ripened pod, and the overwhelming majority of them land somewhere that the right fungus is not, and they sit there in their armored shells, unopened, unfed, and they die without ever becoming anything. Across orchids generally, the rule of thumb biologists fall back on is brutal: fewer than one seed in a hundred ever reaches a seedling that survives its first season. For lady's slippers specifically the picture is, if anything, harsher and slower, and we have it from the ground. When researchers buried mesh packets of *C. calceolus* seed in Danish soil and dug them up year after year, germination showed up only after four and a half years, and in only one of six buried patches. Five of the six produced nothing at all. The dust is cheap, and the plant spends it like the wind: scatter enough, and a few specks will drift down onto exactly the right patch of fungal ground. The rest is loss, planned loss, loss built into the design.

A lady's slipper seed is not a small plant. It is a wager. The flower ships its young out unfinished, unfed, and unable to save themselves, and it does so on purpose, gambling everything on the chance that some of them will fall where a partner is waiting underground to do for them what they cannot do for themselves. Most of that dust is a bet the plant knows it will lose. It scatters anyway. It has been scattering anyway for thirty million years.

There is an irony folded into the chapter's title that I should be plain about. The dust does not dream of being a flower. It cannot dream, or decide, or strive toward anything; it is an inert mote with a clump of cells and a few drops of oil, and left alone it would lie in the litter forever without a flicker of intent. If it becomes a flower at all, it is in effect *dreamed into being* by something outside itself, by the fungus and the soil that find it and feed it. The ambition, such as it is, lives in the partner. The seed only waits to be wanted.

The childhood no one ever sees

Even Charles Darwin, who studied orchids more closely and more lovingly than almost anyone before him, never got to the bottom of this one. When he wrote his great book on the contrivances by which orchids are fertilized, he set the slipper orchids aside on purpose, the genus he excluded from his general scheme, the one he said he knew nothing about. "I leave out of consideration the British species of *Cypripedium*," he admitted in the introduction, "of which I know nothing." It is a striking thing to find in Darwin, who knew so much: an open confession of a blind spot. He returned to the slippers later in the same book, puzzling at the pouch and the way an insect must pass through it, but the deeper life, the part that happens before the flower, stayed shut to him. He was watching the bloom and the bee, as we all do. The childhood underground, the dust that cannot feed itself, the years of fungal infancy before the first leaf: that was below the reach of even Darwin's attention, conducted out of sight, on a timescale and in a darkness his methods could not follow.

So let me take you down to where Darwin could not go. Say one speck wins. Say it drifts down through the litter and the fungus finds it and breaches the coat and starts to feed it. What happens next is the part of a lady's slipper's life that almost no human has ever seen, because all of it happens underground, in the dark, and takes years.

The fed embryo begins to swell. It does not make a root, and it does not make a leaf. It makes, instead, a small pale knob of tissue, a little tuber of fungus-fed cells, and that structure has its own name: a **protocorm**. The word means "before the body," before the plant proper, and that is exactly what it is, a holding stage, an orchid that is not yet an orchid, a colorless lump living entirely on what the fungus brings it. The protocorm has no green in it anywhere. It cannot photosynthesize and does not try. It sits in the soil, often several centimeters down, swelling slowly, threaded through with fungal filaments that are still its only source of food, and it waits. The thing it is waiting for, the thing it is slowly gathering the resources to do, is to send a single leaf up into the light.

How long that takes, no one can tell you to the year, because the sources genuinely disagree and the plant is buried while it happens. Estimates run from about one year to four. One nineteenth-century observer recorded a green leaf within a year of germination; others, watching more carefully, put the first true aerial leaf in the fourth spring. Either way the arithmetic should stop you: a bean reaches its first leaf in days, and this plant takes one to four years to reach the same milestone, all of it spent as a colorless knob underground, fed by a partner it never chose. The waiting is not a virtue here. It is forced. A seed with no lunch has no other way to become a plant, so it does the only thing its condition allows: sit in the dark and slowly accept what the fungus gives until it has banked enough to risk a leaf.

And then, one spring, the protocorm has finally banked enough. It pushes a shoot up through the soil, and for the first time in its life the orchid breaks the surface and unrolls a green leaf into the light.

The long climb to a flower

You might think that first leaf is the finish line. It is barely the start.

Follow the timeline as a ladder and the scale of it lands. A seed sown into the ground takes years just to germinate. Then one to four more years as a buried protocorm before it raises that first green leaf. And then the climb to a flower, which is measured not in weeks but in more years still. The orchid leafs out, photosynthesizes a little, banks a little of its own carbon at last while very probably still drawing on the fungus underground, dies back, overwinters, comes up a little stronger, and does it again. Season after season, it is building a body and a root system and a reserve large enough to afford the staggering expense of a flower, and it cannot afford one for a long, long time.

How long depends on which lady's slipper you are watching. For the pink lady's slipper of the first chapter, *C. acaule,* the figure usually given is somewhere between six and ten years from seed to the first bloom, out there in the wild on acid ground. For the showy lady's slipper, *C. reginae,* the most carefully documented of any of them, the old field studies put first flowering at fourteen to sixteen years. Fourteen to sixteen years. A plant that germinated the spring a child was born might first bloom around the time that child is finishing high school, and most of those years happened where no one could have watched them.

There is a way to cheat all this, and it is worth knowing about precisely because of what it reveals. In a laboratory, growers can raise lady's slippers from seed without the fungus at all, on a sterile jelly of sugars and nutrients in a sealed flask, spoon-feeding the embryo the carbon a fungus would otherwise have supplied. Do that, and the timeline shortens. A wild plant that needs the better part of a decade can reach flowering size in something closer to five or six years in cultivation. Much of that gap is the fungus, or rather the long search for a partner, the odds and the waiting and the years a wild seed spends either failing to find a fungus or eking out a living on what one slowly provides. The flask removes more than that, of course: no browsing deer, no competing roots, no drought, optimized light and food. But take away the need to find the fungus, hand the embryo its groceries on a sterile spoon, and you shave real years off the front of the plant's life. Which tells you, better than almost any other single fact, just how much of a lady's slipper's childhood is spent on the partnership we have not yet properly met. Most of it.

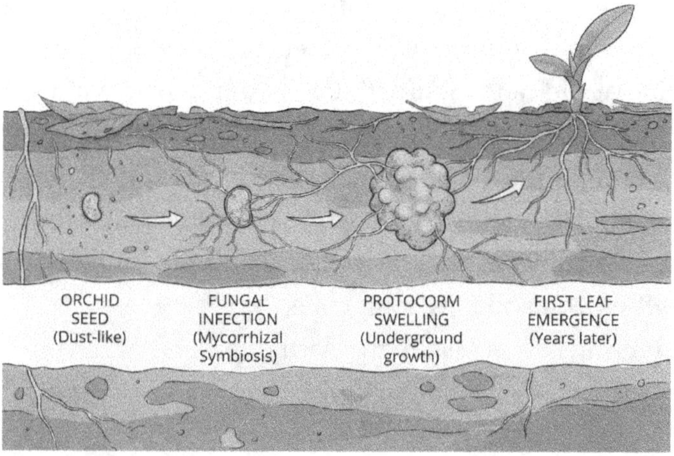

The hidden childhood. A dust seed is found and opened by a fungus, swells into a pale protocorm fed underground, and only years later raises its first green leaf into the light.

What the waiting is for

The naturalists who loved this flower mostly met it in bloom, in the brief weeks when it is impossible to miss, and they wrote about the bloom. Thoreau, listening for the turn of the year in his Concord woods, used the lady's slipper as a clock: "The first of June, when the lady's-slipper and the wild pink have come out in sunny places on the hillsides, then the summer is begun according to the clock of the seasons." It is a lovely line, and it is true, and it is also looking at the very top of an iceberg. The flower that tells Thoreau summer has begun is the visible tip of a life that has spent the better part of a decade, or more, getting to the point where it could afford that flower at all, almost none of it aboveground, almost none of it ever seen. The bloom is not the plant's beginning. It is something the plant saved up for, in the dark, for years.

That is the part I most want you to carry out of this chapter, the next time you are out in the woods in early summer and you find one of these in bloom. The flower in front of you is the least of it. Under your boots, in the cool dark you cannot see into, this plant once was a speck of dust with no food of its own, a wager the parent flower scattered by the thousand and mostly lost; and the one speck that became this plant did so only because something found it underground and fed it through years of waiting before it ever raised a leaf. The patience you are looking at did not begin with the bloom. It began at the very start, in the seed, in the dark, long before there was anything green to see.

Tip the rest of the dust off your palm now. Let the late-summer air take it. Most of those specks are going nowhere, and a few of them, if the ground is right and the partner is waiting, will still be becoming flowers when you and I are long gone.

Which brings us, at last, to the partner.

The Partner Underground

The most important living thing in this book has no flower. It has never had one and never will. It is not pink, not yellow, not built like a slipper or a face or anything you would kneel to admire. It has no leaves, no stem, no roots in the sense you mean the word. It is not even, strictly speaking, a plant. And without it, not one of the twelve orchids these pages are about could exist at all.

I have spent years among fungi, hunting them in the woods and growing them on logs and in a bed of damp wood chips behind the house, and I still find this one hard to hold in my mind, because the orchid is so insistently visible and the partner is so completely not. I am no laboratory mycologist. What I know of fungi I learned the slow way, with my hands in the sawdust. So let me put the partner onstage the way I have actually come to know it.

Split open one of those logs after a season and the wood inside is laced with white: a soft webbing of threads running through the grain in every direction, feeding on the wood and slowly taking it apart. Lift a flap of the chip bed and the same pale web is there in the dark, threaded through the chips, knitting them into a mat you can peel up like felt. That is what a fungus mostly is. Not the mushroom, which is only the brief fruit it sends up to scatter spores, but the hidden net of thread itself, patient and enormous, almost never seen.

But the orchid does something with a fungus that I, for all my growing, have never managed to do. It takes the fungus inside itself, down into its own living cells, and there is no raising a thing like that in a bucket. You cannot see it the way I see the web in a split log; it is far too small, and it happens inside the root.

It takes a microscope, and a root from a plant that has died or been salvaged, because you do not dig these up to satisfy your curiosity. Shave that root thin, thin enough that light passes through, and lay it under the lens. The root cells come up big and squarish and orderly, like rooms along a hallway. Inside some of those rooms, where a normal plant cell would hold only its own quiet machinery, is something that does not belong to the plant at all. A fine thread, finer than anything the orchid could grow, has come in from outside and wound itself into a dense little coil, a knot of living filament packed into the cell like yarn stuffed into a drawer. There is another in the next cell. And the next. The orchid did not make these. Something else did, something that came in from the soil and moved into the orchid's own body, room by room.

That something is the partner. And those coils are the whole story.

What the partner is

The partner is a fungus. A soil fungus, to be exact, one of the unremarkable many that live their lives in the top few inches of the forest floor, in the dark, in the damp, threading through dead leaves and mineral grit and breaking things down. You have walked over them your whole life. Every time you have crossed a lawn or a woodland path or a fen boardwalk, there were miles of fungal thread underfoot, doing the slow patient work that fungi do, and you never gave a single one of them a thought. They have no flowers to catch your eye. That is rather the point of this chapter.

Most of our native lady's slippers, when you take that thin slice of root and read it carefully, turn out to keep company with a narrow group of these soil fungi from a single family, the Tulasnellaceae. Say the family name once and you can let it go; what matters is the kind of thing it names. These are the humble ground-dwellers, the lineage botanists sometimes lump under the old workaday label "rhizoctonia," the fungi that decay litter and live in soil and ask nothing of anyone's attention. They are not the glamorous fungi. They make no mushroom you would photograph. And it is to these, of all the organisms on Earth, that the lady's slipper has handed its entire existence.

The relationship has a name, and it is one of the best names in biology, because it is built straight out of the two things involved. In the eighteen-eighties a German botanist named Albert Bernhard Frank was puzzling over the way certain tree roots seemed to be wrapped and infiltrated by fungal threads, and he needed a word for it, and the word he made was honest and plain. He took the Greek *mukēs*, fungus, and *rhiza*, root, and set them side by side: fungus-root. **Mycorrhiza**. That is all it means. Not a metaphor, not a theory, just a description of the literal thing under the microscope — a fungus and a root grown into each other until you cannot say where one stops and the other starts. Frank was describing trees. But the word he made turns out to fit the lady's slipper better than almost anything alive, because in the orchid the fungus does not merely wrap the root. It goes inside the cells.

Albert Bernhard Frank (1839–1900), the German botanist who in 1885 coined the word mycorrhiza *for the partnership of root and fungus.*

How the bond actually works

Here is what those coils are doing, and it is worth slowing down for, because this is the most intimate arrangement in the whole genus.

The fungal thread grows up to the orchid's root and then through its outer wall, into the living cells, and once inside, it loops and folds back on itself again and again until it has packed the cell with a tight tangle of filament. Botanists who study orchids have a particular word for one of these intracellular coils. They call it a **peloton**, from the French for a little ball or skein of thread, and the word is exact: it looks like nothing so much as a ball of wound yarn sitting inside a glass box. A single infected root cell can hold one of these living skeins; a colonized root holds them by the hundred.

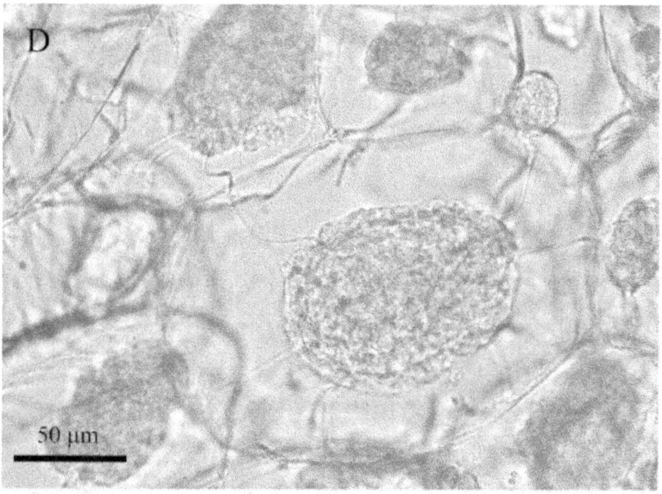

A fungal peloton: a dense coil of fungal threads packed inside a single orchid root cell, where the exchange happens. The orchid digests these coils and the fungus rebuilds them. (Shown here in Pleione.*)*

And then the orchid does something that should stop you. It eats the coil.

Not the whole fungus, and not the thread coming in from the soil, but the peloton itself, the part the fungus has pushed inside the cell. The orchid's cell digests it, breaks the wound filament down from the inside, and takes up much of what it carried: carbon, the sugars that are the fuel of being alive, and minerals the fungus gathered from the soil and the litter. Across that living seam, in the dark, nutrients pass from fungus into orchid. The skein is consumed. And then, remarkably, the fungus grows a fresh one. New threads come in, coil up, are digested in their turn, and the cycle runs again, the fungus building and the orchid eating, building and eating, a slow exchange conducted entirely out of sight.

Think about where this is happening. A bee visits a flower and flies off; two plants might touch root-to-root underground; this is neither. This is the fungus *inside the orchid's own cells,* woven into the orchid's tissue at the scale of a single cell, so that the boundary between the two organisms runs through the inside of the plant. Few living things are joined this closely and still stay two. The lady's slipper does not have a partner the way you have a friend across town. It has a partner the way you have the bacteria in your gut — except threaded right into the cells, and except that the orchid would be dead within its first season without it.

The founding bargain

This is not some local trick that lady's slippers happened to invent. It is the deepest fact about the entire orchid family, older than the lady's slippers, older than the flowers, older than almost anything you can name.

Somewhere far back in the history of plants, in the common ancestor of every orchid that would ever live, this habit began. The ancestral orchid made a switch. It gave up the ordinary partnership most land plants keep, with one group of soil fungi, and took up instead with a different group, the lineage that would carry orchids for the rest of their history. That switch happened once, in one ancestor. There are tens of thousands of orchid species now, on nearly every continent, and all of them, without exception, run on a fungus. The pink lady's slipper in a Maine pine wood and a tropical orchid clinging to a branch in a rainforest and the little green weeds you would step on without seeing are all keeping the same ancient bargain their shared ancestor struck.

So when you find a lady's slipper, you are not looking at a plant that happens to need a fungus. The slipper genus is barely thirty million years old, a recent chapter in a far older story, and it inherited this dependence whole, from the very root of the orchid family. The flower is the new part. The fungus is the old part. The bargain came first.

It is not just a childhood

In the last chapter I left you with a seed that could not feed itself, a dust mote that lived only because a fungus found it in the dark and fed it through years of buried infancy before it ever raised a leaf. That much you already have. Here is the part that surprised even the people who study these plants, and that changes the whole shape of the story.

The dependence does not end when the orchid greens up.

You would think it would. The logic seems obvious. The seedling is helpless because it has no leaf and cannot make its own food, so it eats the fungus; once it finally pushes a green leaf into the light and starts catching the sun, surely it grows up, weans itself, and lives like any other plant, paying its own way. That is the tidy story, and the textbooks told it for a long time, and it is wrong. The green adult lady's slipper does not let go of the fungus. It keeps drawing carbon from underground even after it has leaves of its own, running partly on sunlight and partly, still, on its partner. A plant that lives this double way, half by photosynthesis and half by feeding on a fungus, is called a **mixotroph** — a mixed feeder, and that is what an adult lady's slipper appears to be. The mycoheterotrophy of the seedling, the pure fungus-eating of the years underground, does not switch off at the first leaf. It eases into something lifelong.

We have not measured this directly in most of our native lady's slippers. To measure it you have to read the carbon itself, to trace the particular signature of fungus-derived carbon in the orchid's tissue against the signature of carbon a plant fixes from the air, and that careful isotope work has been done, in the wild, for exactly one native North American species. It is the brownie, the clustered lady's slipper (*Cypripedium fasciculatum*), a small dark western orchid I will come back to in a moment. In green, leafy, photosynthesizing adults of that species, researchers found that a little over forty percent of the carbon in the plant had come not from the air but from the fungus. More than forty percent, in a grown plant with perfectly good leaves. An orchid cousin in Asia, measured the same way, came out at about the same. The adult lady's slipper, as far as we can yet see, never fully leaves home.

So the dependence the seed showed you was not a phase. It was the visible beginning of a thing that runs the whole length of the life.

The pink one's particular faith

Let me bring this back to the orchid you have known longest in this book, because its version of the bargain is the loveliest piece of mechanism I know.

Remember the pink lady's slipper (*Cypripedium acaule*) from the first chapter, nodding on its bare scape in the dry acid soil of a pine wood. Remember that I called it a specialist in difficulty, a plant that lives where the ground is too sour and too poor for most things to bother with, and I promised there was a whole story in that. This is the story. When researchers sampled the moccasin flower's roots across sixteen populations in Georgia and read the fungi inside them, they found the genus's usual narrow company, the Tulasnellaceae, the humble soil lineage. A specialist plant keeping to a specialist partnership. And here is why the sour soil matters, why the pink lives, of all places, on the ground nobody else wants.

The acidity is not an obstacle the orchid overcomes. The acidity is a wall it hides behind.

The pink lady's slipper needs soil that is genuinely acid, around pH four to four and a half, well below the comfortable middle where most plants and most soil life thrive. Push the soil any sweeter, above about pH five, and the moccasin flower sickens and its roots rot. For a long time that looked like simple fragility, a plant too delicate for ordinary ground. It is closer to the reverse. In sweeter soil, the soil's own teeming microbial life surges, a riot of bacteria and fungi all competing, and that riot overwhelms the orchid's narrow, particular partnership before it can establish. The one relationship the lady's slipper cannot live without gets swamped. But sour soil holds that riot down. At pH four and a half, most of that competition cannot function, and into that hush the orchid's quiet fungal partner can settle and hold. So the acid ground, as far as we can yet see, is not where the pink survives despite the hardship. It is where its whole life has come to be arranged around the one relationship it has. The plant did not choose the difficult soil;

its lineage was filtered into it, and there, behind that wall, the partner is safe.

Everything about the pink lady's slipper that looks like weakness, the fussiness, the narrowness, the refusal of good soil, is the plant arranging its entire life around keeping faith with something it cannot see.

Plugged into the forest

The partnership is not even just two organisms. Pull back from the single root cell and the single coil, and the lady's slipper turns out to be wired into something much wider.

The clearest look we have at this comes from the showy lady's slipper (*Cypripedium reginae*), the tall white-and-pink queen of wet calcareous fens, growing in western Newfoundland alongside black ash trees. When researchers cataloged the fungi in the soil there, they found that the showy lady's slipper and the black ash were drawing on the same fungal cast, sixty-six kinds of fungus shared between the orchid and the tree. The orchid is not plugged into one partner. It is plugged into a whole underground company of them, the same company the trees around it are using, shared fungal life running through the floor of the fen and connecting the orchid to the forest it lives in.

A calcareous fen with black ash at its margin. In this lime-rich wetland the showy lady's slipper shares its fungal partners with the surrounding trees.

I have to be honest about what that does and does not prove. A shared cast of fungi is not the same as a proven pipeline. Nobody has yet traced a single atom of carbon flowing from a tree, through the shared fungus, into a wild North American lady's slipper. The connection is real; the current through it is not yet measured. The orchid and the ash share their fungal partners, and it is tempting to imagine the tree quietly feeding the orchid down that shared line, but tempting is not the same as shown.

And there is a quieter, sadder thing in that Newfoundland fen. The black ash, the tree the showy lady's slipper shares its fungi with, is dying across its range, killed off by an introduced beetle, the emerald ash borer, that bores into ash and kills it. Whatever the orchid draws from that shared underground company, one of the largest partners in it is being removed from the forest. The web the lady's slipper lives inside has a hole opening in it.

A few that wandered

I have been telling you the genus norm, the narrow faithful partnership most of our lady's slippers keep with that one humble soil family. But a couple of them have made stranger arrangements, and before I close I want to set them in front of you, lightly, as coming attractions rather than answers, because they are worth the wait.

The California lady's slipper (*Cypripedium californicum*) lives, as its name does not quite tell you, in a place that is barely habitable: serpentine seeps, wet ground over a strange metal-rich rock that poisons most plants and that cannot move or be replaced. Pinned to one impossible address, the most narrowly placed of all our western slippers. And yet underground, this most particular of plants is the genus's great generalist. Where its cousins keep to one fungal family, the California lady's slipper takes three, in roughly equal measure, an open-handedness no other native lady's slipper shows. The specialist above ground is the generalist below it. There is a whole chapter in why.

A serpentine seep: cold water sliding over metal-rich rock, ground too poor for most plants. It is the narrow world the California lady's slipper is pinned to.

And the brownie, the small dark clustered lady's slipper I mentioned a moment ago, the only one whose adult carbon we have actually measured, has wandered furthest of all. Its partners are not the humble soil fungi of the genus norm at all. They are forest fungi, the kind that ordinarily live in partnership with the great conifers, fed by Douglas-fir and the trees around it. The brownie has plugged itself into the conifers' own underground network and drawn its forty percent of carbon down that line, a green orchid quietly turning into something that lives partly by tapping the forest. What exactly it has become, and even which fungi it truly keeps, is genuinely argued over by the people who study it, and I am not going to settle that here. I only want you to know that the bargain, faithful and narrow in most of the genus, can be bent into shapes this strange. We will go west and meet these two properly later.

How little we have met the partner

And for all that I have told you about the partner, the wanderers included, there is one more thing, and it is the thing that finally sets up everything this chapter has been walking toward. We barely know this fungus.

For decades, when scientists went looking for the fungi inside orchid roots, the very tools they used were nearly blind to the one that matters most. The standard molecular methods, the genetic probes meant to read whatever fungus was present, badly undercounted the Tulasnellaceae, the lady's slipper's own central partner. The fungus was right there in the root, in the coils, doing the work of keeping the plant alive, and the instruments kept missing it. It took special methods, primers designed on purpose to see this overlooked family, before the partner came clearly into view at all. We were looking straight at the thing the orchid cannot live without, and for years we could hardly see it.

And even now, vast reaches of the lady's slipper's world have simply never been examined. Take the pink one again. We know its fungal partner from those Georgia populations and almost nowhere else. Across the great northern body of its range, the boreal forests of Canada, the woods of New England, the country around the Great Lakes, no one has ever sampled which fungus is feeding the moccasin flowers that grow there by the thousand. We owe the existence of these orchids, every one of them, to an organism we are only now learning how to name, in places we have not yet looked.

What to call this

So now you have the real picture, and I can finally hand you the word I have been holding back since the very first chapter.

Let me lay the picture out one more time, plainly. A lady's slipper begins as a speck of dust that cannot feed itself and would die where it fell if a fungus did not find it. The fungus opens it, feeds it, threads itself into the orchid's own cells and stays there, building coils the orchid digests and rebuilding them again, for years underground and then on through the green adult life, so that even a grown, leafy, sunlit lady's slipper is still drawing part of its living from a partner woven into its tissue. The orchid did not choose this partner. It cannot see it. It found it by sheer chance, one speck of dust in thousands drifting down onto the one patch of ground where the right fungus waited.

And it gives, as far as we can honestly tell, very little back. The seedling takes carbon and offers nothing in return, no green tissue, nothing to trade. Whether the adult ever truly repays the fungus is one of the real open questions in the science: the most careful recent reading calls the orchid a carbon exchanger with some return flow, a hedge rather than a verdict, and other researchers have asked the darker question of whether the fungus, on a dying orchid, might turn and feed on it. Mutualism, the tidy textbook word for two partners who both come out ahead, is a word the data do not fully earn here. We do not actually know that the books are balanced. We may never.

And here is the thing. That uncertainty does not weaken the story. It is what makes the honest word for it the right one.

Because look at what the lady's slipper has actually done. It has staked its entire existence on a partner it cannot see, did not choose, and cannot obviously repay. It has put its life, from its first hours and on through every green year that follows, into the keeping of something it has no way to hold onto, no way to verify, no way to make stay. And the partnership holds. It has held, in this lineage, through this exact arrangement, for some thirty million years. We tend to call that kind of dependence fragility. I told you back in the first chapter that I would try to convince you it might be better to call it trust. This is where I make good on it. Not the soft kind of trust, the comfortable kind between equals who can each walk away. The other kind. The kind where you have no leverage and no proof and no fallback, where you have given yourself over completely to a partner you cannot account for, and the giving-over is the whole of how you live. The lady's slipper is not fragile because it depends. It is honest about depending. It wears its dependence at the root, where we can finally see it.

And the reason that should land, the reason it is not just a pretty thing to say about a flower, is that we are not so different. We prize independence and read need as weakness, but nothing alive is actually independent. We too are fed, held, and kept going by partners we did not choose and mostly never see, threaded so far into our own bodies and lives that we have stopped noticing them at all, even if we like to imagine we stand on our own. The lady's slipper is not the strange exception to how living things get by. It is the plain rule, made visible, grown where we can finally look at it.

Down where you cannot see

I have a habit, when I find a lady's slipper now, that I did not have when I was younger. I still look at the flower, of course. You would have to be made of stone not to. But after a while my eye goes down, past the pouch and the leaves, to the few inches of ordinary leaf litter and dark soil at the base of the plant, the part that holds no spectacle at all.

That is where the partner is. Somewhere in that handful of unremarkable ground, finer than thread and impossible to find, is the fungus this whole flower is built on, doing right now the slow work it has been doing since before the pouch opened: holding the coils, passing the carbon, keeping the orchid. You cannot see it. You will never see it standing up. But it is there, and the flower above it is, in the most complete sense, its guest, and has been since the first dark hours of its life.

The orchid took years to raise that bloom, and it will be gone in a week or two. The partnership underneath it does not keep that schedule. It was here before the flower and it will be here after, down in the dark, patient on a timescale that does not bother with summers, waiting, the way it has always waited.

The Long Wait

In 1943, the botanist Curtis published the result of a kind of patience we no longer have much room for. He had been watching showy lady's slippers grow. Not for a season, not for a few summers, but long enough to answer a question almost nobody had thought worth the trouble of answering, because answering it took most of a working life: how long does it take, from a single dust seed, for one of these plants to make its first flower?

Fourteen to sixteen years.

Let that number sit for a moment before we do anything with it. The showy lady's slipper, *Cypripedium reginae,* drops its seed in autumn. The seed germinates a few centimeters (an inch or two) down in the dark, where, as we saw two chapters ago, it has no leaves and no chlorophyll and survives entirely on a fungus it cannot live without. It is a protocorm first, a tiny pale knob of tissue with no resemblance to a flower, eating coil after coil of its underground partner. Around year three, the first green shoot breaks the surface. And then it waits. Year four, year five, year eight, year ten, a small plant photosynthesizing quietly in the leaf litter, putting up a leaf or two each spring and pulling everything back down each fall, building something underground we cannot see. Somewhere in the fifteenth year, give or take, it finally spends what it has saved on a flower.

A person who plants a *reginae* seed and wants to see it bloom from their own labor had better be in good health. By the time the flower opens, a child born the same autumn is finishing high school. The plant that began as a speck of dust too small to see clearly has spent the better part of two decades becoming, finally, the thing we recognize. We show up for the last act and call it the play.

A showy lady's slipper standing nearly waist-high. The flower is the brief, costly tip of a plant that needed well over a decade to reach its first bloom.

The patience champion, and the rest

The showy lady's slipper is the slowest of the well-documented species, which is part of why botanists keep coming back to it. The pink lady's slipper, *Cypripedium acaule*, and the yellow, *C. parviflorum*, are usually put at six to ten years from seed to first flower. Faster, but only in the way a long sentence is shorter than a paragraph. Across the genus the early calendar runs more or less the same: germination underground, the first green leaf typically sometime in year two to four, and then years of leaf-without-flower before there is enough in reserve to bloom.

I should say plainly that there is a shortcut, because we met it in the chapter on the seed and I don't want you to think the whole genus is locked into this. Grown in a flask, fed sugar directly, spared the search for a fungal partner and the long underground apprenticeship, a lab plant can reach flowering size in about five or six years. That is the fast track, and it is fast only by the standards of a plant that otherwise measures its childhood against the lifespan of a dog or a marriage. In the wild, with no such help, the climb takes its full length. There is no hurrying it.

What are those years for? Not nothing. A lady's slipper that puts up one small leaf a summer is not idling. Every season it surfaces, it photosynthesizes a little, and every fall it withdraws what it can underground into a rhizome, a thick horizontal stem that runs just below the surface and serves as the plant's pantry and its anchor and its long memory all at once. The slow years are accumulation. A reserve builds up gram by gram, season by season, against the enormous one-time cost of a flower, and against the leaner years that will come. A flower is expensive: the lip alone, that inflated pouch we will spend most of Part III admiring, is a structure built to manipulate an insect, and structures that complex are not cheap to build. No bloom appears until the reserve is large enough to make one and still outlast the cost of making it. From a leaf or two of sun and the fungus underground, that takes roughly fifteen years.

Here is the thing worth holding onto. We tend to file slow growth under hardship, as if the plant were straining toward the flower and merely failing to get there quickly. That is the wrong picture. The plant is not behind schedule. This *is* the schedule. The years of leaf, the seasons of nothing aboveground but a green tongue in the litter, are not the prelude to the lady's slipper's life. They are most of its life.

And to see how brief, and how expensive, we have to talk about how long these plants actually live, which turns out to depend entirely on what you mean by "one plant."

Two kinds of individual

Walk up on a big lady's slipper clump in June and you might count a dozen stems, or twenty, each with its own leaves, some flowering, some not. The honest question is: are you looking at twenty plants, or one?

The answer is both, and the difference between the two answers is the most important idea in this chapter.

Botanists who work on plants like this keep two words for it. A **ramet** is a single stem, a single shoot, one of the things you actually see and could touch in a given year. A **genet** is the whole genetic individual, the original plant and everything that has grown from it as a clone, the entire clump that all those stems belong to. One genet, many ramets. The clonal habit of *Cypripedium* means a single seed that germinated long ago can spread sideways underground along its rhizome into a slowly widening colony of shoots, all of them genetically one plant, the way a stand of aspen on a hillside can be a single tree pretending to be a forest. The ecologist who gave us those two words, John Harper, established them in 1977, trying to fix exactly this confusion: that when we count plants, we often don't know whether we're counting individuals or counting parts.

This is not a technicality. It changes the number you put on a lady's slipper's age by a full order of magnitude. It also changes what you are even counting. A field survey that reports a hundred lady's slippers might be reporting a hundred separate genetic individuals, the product of a hundred lucky seeds, a thriving population. Or it might be reporting a few old genets that have each spread into twenty or thirty stems, which is a very different and much more fragile thing: a handful of individuals wearing the costume of a crowd. The flowers look the same either way. You cannot tell the two situations apart by counting blooms. You have to know which stems belong to which buried plant, and that knowledge, like everything else about this orchid, has to be dug for slowly.

Take the ramet first, the stem. How long does any one stem last? We know this with unusual confidence for *reginae,* because two ecologists, Marc Kéry and Katharine Gregg, did the slow work of marking individual plants and returning to them year after year, recording who came back and who didn't. They found that a given plant survives from one year to the next about ninety-five or ninety-six percent of the time. That sounds like near-immortality until you do the arithmetic on what a ninety-five percent annual survival rate actually buys you over time, and it works out to a ramet living on the order of twenty to fifty years — the high end exceptional. A good run, for a stem. Longer than most wildflowers manage as whole organisms.

Now the genet. The clump.

Many stems flowering from one spot. A patch like this can be a single genetic individual, spreading slowly underground and outliving any one stem by generations.

Century-scale

The strongest numbers come from across the Atlantic, from the yellow lady's slipper of Europe, *Cypripedium calceolus* — a different species from our own yellow, though close kin, and studied longer and harder than anyone has studied any *Cypripedium* here. When a researcher named Nicolè and colleagues modeled the lifespan of *calceolus* genets in 2005, the numbers they got back were not in decades. Not decades. They were a hundred and ten to three hundred and fifty years. An earlier study, by Shefferson and colleagues in 2001, put the same species' genets somewhere between twenty-five and a hundred. The two estimates disagree by several-fold, and that gap is the honest part: these are models, built from survival rates and growth, and they are only as firm as their assumptions. But notice that even the lower study's cautious figure, twenty-five to a hundred years, describes a plant that can outlast the person who finds it.

I want to be careful here, because this is exactly the kind of fact that gets repeated until it loses its edges, and the precision matters. Century-scale longevity belongs to the genet, the clump, the genetic individual. It does not belong to the stem. No single shoot of a lady's slipper lives three hundred years. What lives three hundred years is the buried, branching, slowly creeping thing underneath, the original plant, replacing its visible stems the way you replace the cells of your own skin, so that the individual persists while every part you can see is young.

Two more honesties. The biggest numbers are European and they are modeled, inferred from survival rates and growth, not read off a single plant that someone watched for three centuries. Nobody has lived long enough to do that. And no marked North American lady's slipper has been tracked anywhere near that long, simply because the studies haven't been running for centuries. So when I tell you our showy and our pink and our yellow are very likely century-scale plants, I'm reasoning from their European sister and from what we can see of their own slow clonal spread, not from a North American clump with three hundred candles on it.

But the inference is sound, and it shows up at home in a smaller, quieter way. The small white lady's slipper, *Cypripedium candidum,* a prairie plant, also forms long-lived clones. Some clumps in the Midwest and in Manitoba are believed to be many decades old, persisting in remnant prairie that has had nearly everything else stripped from it. A lady's slipper colony, in other words, can be the oldest living thing in a place and look like nothing in particular.

There is a cost buried in all that endurance, and it is the same slowness wearing its other face. The thing that makes a single genet nearly unkillable is exactly the thing that makes the species almost impossible to replace. Fewer than one in a hundred viable seeds ever finds its fungus and lives to become a seedling, so new individuals arrive rarely, and in some places they stop arriving at all. A colony can drift into being all elders, every clump many decades old and flowering faithfully, with nothing young coming up behind it. The plants are not dying. They are aging out, holding the ground beautifully and growing no replacements, a crowd of the nearly immortal with no next generation in it.

So here is the reframe, and I've made you wait for it the way the plant makes the bee wait, the way the seed makes the season wait. When you find a lady's slipper, the stem in your hand is the youngest part of it. We arrive, we admire the flower, we photograph the flower, and the flower is a few months old. The plant is potentially older than the country we're standing in. We have been looking at the most temporary, most replaceable, most recent feature of a very old, very slow individual, and calling that the plant. The flower is not the life. The flower is the part of the life that happens to be visible this week.

Time is the medium this plant works in. It is also, as we will see, the thing about it you cannot move.

The years it can skip

There is one more move in the lady's slipper's long game, and it undoes any tidy story you might have just built about a stem coming up faithfully every spring for its twenty to fifty years.

It doesn't have to come up.

A lady's slipper can go a whole year, sometimes more than one, without putting up anything aboveground at all. No stem, no leaf, no flower, nothing. The plant has not died. It has gone back to living entirely underground, on the fungus, the way it lived as a protocorm before it ever surfaced, and then a year or two later it returns as if nothing happened. Botanists call this **vegetative dormancy**, and Kéry and Gregg documented it in the very same showy lady's slippers they were marking, watched tagged plants that were unmistakably alive simply decline to appear, then reappear seasons later. It is documented across the genus, and the demographer Richard Shefferson has built much of a career on it, on the unsettling fact that a plant can be absent and present at the same time.

Think about what this does to the act of looking. You learn a lady's slipper's spot. You go back the next June and it isn't there. The natural conclusion is that something killed it, that the place is poorer than it was. But a lady's slipper that fails to show up may be perfectly alive, riding out the year underground, putting up no stems or leaves at all until conditions improve. The plant you cannot find is not necessarily a plant you have lost.

It also makes the demography genuinely hard, which is worth sitting with for a second. If a plant can be alive and invisible, then how do you ever know it died? You don't, not for years. A tagged plant that misses a season might be dormant or might be gone, and the only way to tell the difference is to keep coming back long enough to give it every chance to reappear. A plant absent for one year is probably resting. A plant absent for five is probably dead, but even that is a judgment call, an act of patience pitted against the plant's own. The researchers who measured these lives had to learn to wait the orchid out.

This is the same patience we started with, only now it's not stacked at the front of the life as a single long childhood. It's distributed through the whole life, available in any given year. The waiting is not a phase the plant passes through on its way to becoming a flower; the waiting is built into how it stays alive, a capacity it can draw on at any age.

A numbered tag staked beside a wild plant. Years of patient tagging are how researchers measured these orchids' long lives, and their habit of vanishing underground for a season.

How we know

I have been throwing these timescales at you, fourteen to sixteen years, twenty to fifty, more than a century, and it's worth stopping to ask the question that ought to come right after wonder: how could anyone possibly know this?

Not by guessing. Not by cutting one open and counting rings, because there are no rings to count. The only way to learn how long a slow plant lives is to mark individual plants, give each one a small numbered tag, write down where it stands, and then go back. Spring after spring. Year after year. Decade after decade. You record which plants flowered, which only leafed, which vanished, which came back from the dead the botanists had quietly assumed they were. This is called mark-recapture demography, and it is one of the least glamorous and most honest things a person can do with their attention. Kéry and Gregg's survival numbers, Shefferson's dormancy work, Curtis tracing the seed-to-flower climb back in the nineteen-forties, every one of those figures is somebody's years.

There is a humbling symmetry in it. To measure the patience of a plant that takes fifteen years to bloom and may live for centuries, a human being has to spend a serious, irreplaceable chunk of a single human life — kneeling in the same patch of ground with a notebook, season after season. The showy lady's slipper is the genus's model organism for exactly this reason: it became the plant we understand best because someone was willing to watch it long enough. You cannot rush this knowledge. The plant won't let you. You can only match a little of your time against a lot of its time, and write down what you see.

And the watcher will not outlast the watched. That is the strange dignity of this kind of science. A demographer who marks a young lady's slipper and follows it faithfully for thirty years will likely retire, or die, before the genet they were studying is anywhere near old. The plant outlives its own biographers. The data get handed on, the tags rust, a new researcher inherits the plot, and the orchid goes on being slow, indifferent to how many human attention spans it has now quietly outrun.

Which brings us, finally, to the trowel.

Why you cannot dig one up

People try to take these home. It is one of the most natural impulses in the world. You find a lady's slipper blooming in the woods, the most beautiful wild thing you have ever stood next to, and some part of you wants it in your own garden, by your own door, where you can see it every June. So you bring a shovel.

It almost never works. Transplanting a wild lady's slipper fails so reliably, across the whole genus, that the failure is itself one of the best-documented facts about these plants. And the pink lady's slipper is the worst of all: the survival rate for a wild *acaule* moved by a hopeful gardener is about five percent. Nineteen out of twenty are dead within a few years. Reintroductions run by people who actually know what they're doing fail for the same reason; the Boston University botanist Richard Primack has documented how stubbornly these plants resist being re-established around Boston. This is not a problem of doing it carefully enough. It is a problem of what the plant is.

Here is the mechanism, and once you see it, the five percent stops being mysterious.

When you dig up a lady's slipper, the plant usually survives the first year. It comes up, it may even flower, and the gardener is encouraged. But that first year is borrowed. The plant is living on what it had already stored in its rhizome, the savings account underground, the same reserves it spent fifteen years filling. What it is not doing is making more, because the thing that fed it for its entire life, the fungus, is gone. Cut off in the original soil. Left behind in the woods. The fungal partnership we spent a whole chapter on does not travel in the root ball, and it does not re-establish in your garden bed, where the soil chemistry is wrong, where (for the pink lady's slipper especially) the ground isn't acid enough, where the particular fungal partners simply aren't present. So the second year is leaner, and the third leaner still, and the reserves run down with nothing replacing them, and somewhere in there the plant quietly starves in the most beautiful soil you could give it.

You did not move the plant. You moved the visible part of the plant and severed it from the partner that keeps it alive, and the visible part coasted on its savings until the savings ran out.

And now put this together with everything else in this chapter. What were you actually trying to dig up? Not a flower. The flower is months old; the flower was never the point. You were trying to move a relationship that took more than a decade to build underground, a partnership between a plant and a fungus that cannot be transplanted because it isn't a thing, it's an agreement between two things, and you can only carry one of them in the bucket. And if it was an old clump, you were trying to move a century. A hundred years, or three, of slow sideways growth, of stems raised and dropped, of years spent dormant underground, of patience compounded season on season into a single genet older than the house you wanted to plant it next to.

You cannot dig up that much time. There is no tool for it. The shovel reaches the roots and stops there, and everything that actually matters, the partner you can't see and the years you can't carry, stays in the ground.

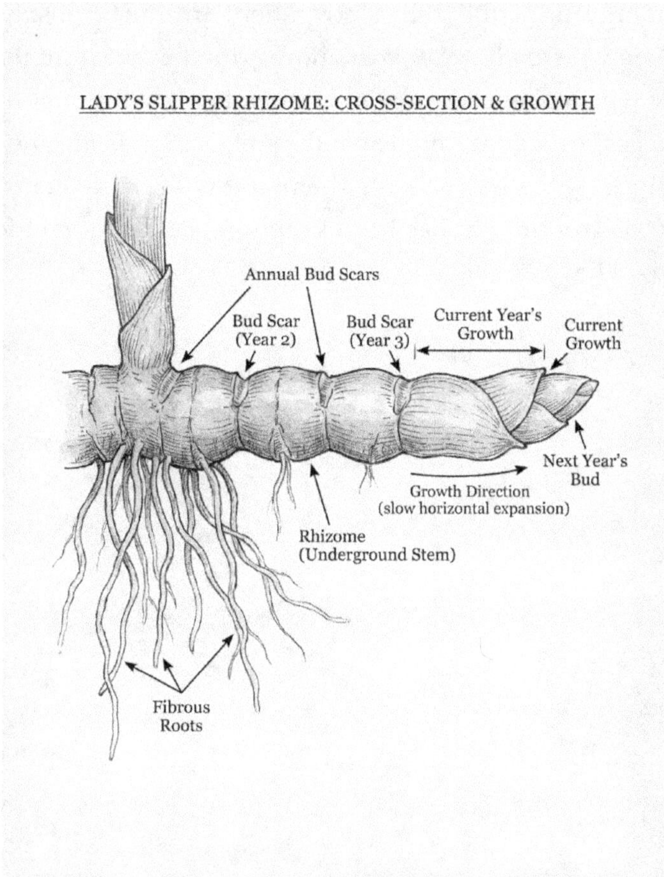

A cross-section of the rhizome, the slow horizontal underground stem. Its stored reserves pay for each year's growth and refill only slowly once spent.

What stays in the ground

So the lady's slipper is, of all things, a lesson in scale we are badly equipped to perceive. We are built to notice the flower. We show up at flower speed, with flower-length attention, and we mistake the part that is keyed to our timescale for the whole organism. The flower is the one feature of this plant fast enough for us to relate to at all. Everything else about it runs on a clock that makes ours look frantic.

The pink lady's slippers I know best grow on a pine ridge near home that I have walked for years. I have no idea how old they are. I never will. But I have stopped thinking of them as flowers I visit. They were there, underground, doing the slow patient arithmetic of staying alive, before I knew the path. Most years now I don't even count the blooms. I just note which ones came up, and remember that the ones that didn't are not necessarily gone, and leave the soil exactly as I found it.

Next June, if the season suits them, a few of those stems will rise again out of ground that has been keeping its oldest secret for longer than anyone has been alive to ask.

The Deception, in Detail

A bee is landing on a yellow lady's slipper, right now, and you have a few seconds to watch.

She is small. Not a queen, not a bumblebee, nothing with the heft and the muscle the first chapter made so much of. She is a sweat bee, a female *Lasioglossum*, a slip of an insect maybe a third the size of the queen who forces her way into the pink lady's slipper down the front. The flower she has chosen is a yellow lady's slipper (*Cypripedium parviflorum*), pouch the color of butter held up to a window, two spiral petals twisting off to either side like loosened ribbon. She works the rim a moment, drawn by the color and a faint sweetness on the air, and then she goes in over the inrolled edge of the opening and drops into the bowl of the pouch.

A female Lasioglossum *sweat bee, one of the small bees a yellow lady's slipper is sized to fit.*

And now she is in trouble, though she does not know it yet. The walls inside are smooth and faintly translucent and offer her nothing to grip. There is no nectar in here. There never was. She came for a meal that the flower only promised, and the way she came in is not a way she can easily go back out.

Watch what she does next.

The same machine, twelve times over

In the first chapter I walked you, slowly and on purpose, through one flower's trick: the pink lady's slipper, its pouch slit down the front, a queen bumblebee shoved in and steered out the back past the stigma and then the anther, dusted with pollen and none the richer. What I did not tell you is that the con is not the pink lady's slipper's private invention. It is the family business. Nearly every one of the twelve native lady's slippers in this book runs some version of it, and the small yellow flower with the sweat bee inside it is running it right now.

The machine has a name. Pollination ecologists call it a **kettle trap**: a bloom shaped into a chamber an insect enters easily and escapes only by a single forced route, the way a fish slips into a fish trap through a funnel it cannot find again from the inside. You already know every part of it from the pink one: the inflated pouch, the inrolled lip that folds shut behind the visitor, the slick walls with no toehold, the lit windows at the back that pull a frightened bee toward the one narrow exit. And you know the masterstroke waiting there, the stigma set before the anther, so the bee leaves another flower's pollen before she picks up this one's. The trap does not just use the bee. It uses her to force the orchid to outbreed.

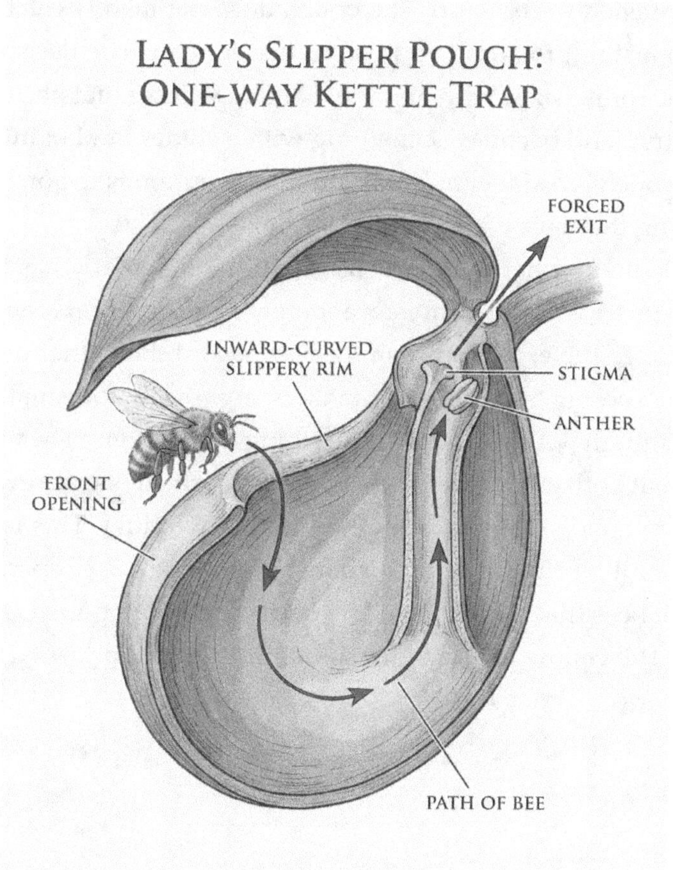

The pouch as a one-way kettle trap. Slick walls and lit rear windows steer the trapped bee to the only exit, past the stigma first and the anther second.

That is the kettle trap, and it is the genus signature. The pink slipper runs it with a queen. The little yellow one is running it, as we watch, with a sweat bee no bigger than a grain of rice.

The trap is really a sieve

Now for the part the first chapter did not give you, the thing that turns a clever flower into something stranger and more exact.

Go back to the two openings, the pink and the yellow, and notice the size mismatch I have been laying down on purpose. A queen bumblebee at the pink. A sweat bee at the yellow. That is not an accident of who happened to show up. The pouch does not merely catch a bee. It selects one, built to admit and use an insect of a particular size and to turn every other size away, and the moment you see that, the trap stops being a trap and becomes something with a colder kind of intelligence in it. A pouch is a **sieve** for bees, and the deception is the orchid's way of running the sieve.

Think about what the right-sized pollinator actually has to be. Big enough to trip the mechanism and force the one-way route. Small enough to fit the exit and brush the stigma and the anther on the way through. Too big and she can't make the passage; too small and she rattles through without touching anything that matters. The window of bodies that both enters correctly *and* leaves correctly is narrow, and the flower's whole architecture is tuned to that window. This is why the books had it wrong for so long about the pink lady's slipper, blaming the small bees that work the yellow ones. The pink's pouch is sized for muscle. The yellow's is sized for something slight. Same machine, two settings.

The pouch is also a sieve. Different lady's slippers are sized for different bees, from a bumblebee queen down to a tiny sweat bee.

And the most beautiful demonstration of the sieve in the whole genus is what happens when you let the same big bee try two different flowers that solve the problem at opposite ends.

The pink lady's slipper, *Cypripedium acaule,* runs its sieve at the **entrance**. You already know how. Its pouch is the only one in the genus that opens not through a round hole but through a long vertical slit down the front, the two edges of that slit rolled inward into a one-way valve. To get in at all, an insect has to push those rolled lips apart by main force, and only a big strong bee, a queen bumblebee fresh out of hibernation, can do it. The slit is the filter, and it filters at the door. Weak need not apply. The flower has decided who its pollinator is before the insect is even inside.

The showy lady's slipper, *Cypripedium reginae,* runs its sieve at the **exit**, and this is the fresh and faintly outrageous one. The showy slipper does not slam its door. Its pouch opens through a generous round hole at the top, wide and welcoming, an easy entrance for almost anybody, queen bumblebee included. She climbs right in. And then she discovers the catch. The two rear exits, the narrow passages beside the column that every lady's slipper makes its visitors squeeze through, are in the showy slipper very small. Researchers measured them: about 6.5 by 3.4 millimeters (roughly a quarter inch by an eighth), a slot that fits a bee of around 4.4 millimeters wide and 3.4 deep and almost nothing larger. A queen bumblebee is far too broad for it. She can get in the front, but she cannot get out the back. So she does the only thing left to her. She turns around and leaves the way she came, out the wide front door, without ever having been steered past the anther, without ever picking up a grain of pollen.

The queen bumblebee walked into the showy lady's slipper and walked out again completely useless to it. She was a visitor and never a vector. The flower let her in, fed her nothing, and let her go, and the whole transaction did the plant no good and no harm. The bees that actually do the work, the ones whose bodies fit that narrow rear slot, are mid-sized: leaf-cutter bees (*Megachile*) chief among them, with a digger bee or two, insects that both enter and exit correctly and carry the orchid's pollen out on their backs the way the design requires. Everyone else who comes is just traffic.

So look at the two flowers side by side. Two lady's slippers, the same problem, opposite solutions. The pink one screens at the front with a slit only a strong bee can force. The showy one throws the front door open and screens at the back with an exit only a mid-sized bee can clear. One says *you must be this strong to enter.* The other says *anyone may enter, but only the right size may leave by the working door.* The trap and its twin. And both of them are doing the same deep thing, which is not catching insects at all. It is choosing them.

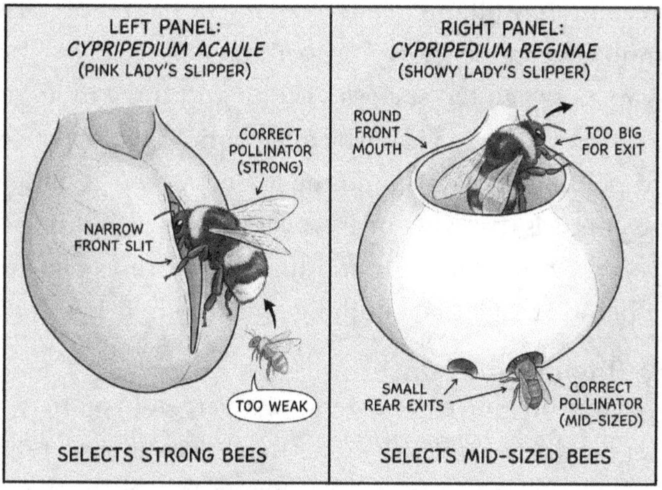

Two ways to sort a pollinator. The pink lady's slipper filters at the entrance, admitting only a bee strong enough to force the front slit; the showy filters at the exit, letting many in but only the right size out the back.

Visited by many, pollinated by few

The sieve is easiest to believe when you stop thinking about machinery and start counting bees, and for the yellow lady's slipper somebody has done exactly that.

Go back to our sweat bee, still working her way toward the light at the back of the yellow pouch. She is one of a crowd. In a Virginia study of the yellow lady's slipper, researchers watched the flowers carefully and tallied the insects that came: ten different kinds of bee, across four families, all of them drawn in by the same yellow promise, all of them crawling into the same pouches. Ten species visiting. That sounds like a popular flower, a well-served one.

Then the researchers checked which of those insects actually left carrying pollen, which ones had been steered through the working passage and loaded on the back the way the flower needs. The answer was three. Not three species. Three individual bees. Two of one kind of little mining bee, one of another, out of the whole milling crowd. Everyone else had come and gone without doing the orchid the slightest good — too big or too small or wrongly shaped to clear the exit and brush the column, walking in and walking out as empty of use as the queen bumblebee in the showy slipper.

That gap, between ten species visiting and three individuals pollinating, *is* the sieve, made visible by arithmetic. The flower is visited by many and pollinated by few, and the fewness is not a failure. It is the point. The pouch is throwing nearly everyone back, holding out for the one body in the crowd that fits and turning the rest of its visitors into a passing parade that costs it nothing and accomplishes nothing.

Why the flower gives nothing

There is a question I have been stepping over, and you may have been waiting for me to trip on it. Why does the flower offer no reward at all?

It would be the simplest thing in the world to put a drop of nectar at the back of that pouch. Most flowers do. Nectar is the ordinary wage a plant pays a pollinator, a little sugar in exchange for the carrying of pollen, and it works: a rewarding flower gets visited often, gets visited eagerly, gets its pollen moved by insects who learn that this kind of bloom pays and come back for more. The lady's slipper pays nothing. Every one of our twelve native species is nectarless, dry at the back, all promise and no delivery. The flowers lure their pollinators with color and scent and the shape of a good meal, and then they welsh. Biologists have a name for the whole strategy, and it is worth setting down plainly: **generalized food deception**, the advertising of a meal the plant has no intention of serving. The first chapter called this a con, and a con it is. But the genus runs it so consistently, across every species, in defiance of how easy it would be to just pay the bee, that the consistency itself demands an explanation. Nobody refuses to pay their workers by accident, twelve times over, for thirty million years.

Here is the explanation, and it is the kind of logic that makes a ruthless thing suddenly make sense.

A bee that gets a reward from a flower does something predictable afterward. She goes to the next flower on the same plant. Why wouldn't she? It paid once; it will likely pay again, and it is right there. A lady's slipper, though, often carries more than one bloom, and two flowers on one plant are the same plant, genetically. If a rewarded bee worked her way down a single stem, flower to flower, she would be carrying that plant's pollen to that plant's own stigmas, fertilizing it with itself. Botanists call that **geitonogamy**, a plant pollinating itself by way of an insect ferrying pollen between its own flowers, and for an orchid that has gone to all this trouble to force outbreeding, it is the quiet enemy. It produces seed, but inbred seed, the very thing the stigma-before-anther sequence was built to prevent.

A flower that gives nothing solves this at a stroke. Our sweat bee, finding the yellow pouch empty, dragging herself out cheated and irritated, will not try the next flower on that plant. She has learned, in one bad visit, that this kind of bloom is a swindle, and she leaves the plant entirely and carries her hard-won load of pollen away to some *other* lady's slipper, possibly yards off, a different individual, before she gets fooled all over again. The deception drives her away from the plant she just pollinated. And so every successful pollination a lady's slipper gets is an outcross, pollen moved between separate plants, fresh genes meeting fresh genes, with the self-pollinating trap of geitonogamy shut tight. The orchid has traded the *quantity* of visits a nectar reward would buy for the *quality* of every visit it does get. It would rather be visited rarely and outbred than visited often and inbred.

The cost is real and the orchid pays it in full. Across the genus, fruit set tends to run somewhere between a twentieth and a quarter of the flowers in a given year, often less. Most lady's slipper blooms, most seasons, are never pollinated at all. They open in their gorgeous deceit and close again having fooled nobody, the bet on the rare right bee simply lost that year. And the plant compensates the only way a slow organism can, by waiting better than anything else in the forest: flowers that stay open and receptive for weeks, stigmas that stay willing far longer than a paid flower would need, pollen that keeps its potency for an unhurried age. The lady's slipper cannot make its pollinator come more often. So it makes itself ready for longer, and holds out, and lets the empty seasons pass. The first chapter put this as a single line, that a few well-aimed deceptions are worth more than a thousand honest payments of nectar. This is the arithmetic underneath it. The flower is buying outcrossed seed at a steep price and refusing, on principle, to buy anything cheaper.

Darwin comes back to the slipper

Somewhere in the middle of all this watching there is a moment I love, and it belongs to the most famous watcher of orchids who ever lived.

When Charles Darwin published his book on how orchids get themselves fertilized, in 1862, he left the lady's slippers out, confessing he knew nothing about them. We saw that two chapters ago. What matters here is what he did next. He went back. In 1877, in the second edition of that same book, Darwin returned to the slippers he had passed over, and this time he worked the machine out. Building on the European yellow lady's slipper (*Cypripedium calceolus*), the Old World cousin of our native yellows, and on reports sent to him of American species, he reasoned through exactly the thing we have been watching the sweat bee do: how an insect is led into the pouch and made to leave by a single forced route, in one door and out another, past the organs that load and receive the pollen. The man who had confessed the genus baffled him came back fifteen years later and read it, the one-way trap and all.

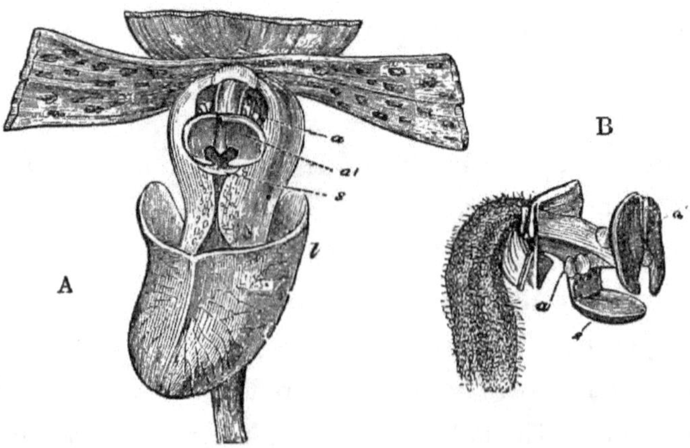

Charles Darwin's diagram of the route a pollinator takes through a lady's slipper, from the 1877 edition of his book on the fertilization of orchids.

I find that quietly encouraging, and not only as biography. The flower that stopped Darwin in 1862 was not unreadable. It was just patient, and it kept its secret until someone watched it long enough and came back. We are still doing that. There are lady's slippers in this book whose pollinators have never been seen and whose fungi have never been named, and somebody, someday, will go back to them the way Darwin went back to the slipper, and finally read the machine.

The orchid that walked away

I have been telling you, this whole chapter, that the deception is brilliant. I believe it is. But brilliance has a hidden condition buried in it, and there is one lady's slipper that found the condition and called the orchid's bluff.

The whole apparatus, the trap, the sieve, the rewardless lure, the forced outcrossing, only pays in a world with bees in it. Every bit of the cleverness assumes a pollinator will eventually come, the right-sized dupe out of the milling crowd, often enough across the long-lived years to set some seed. Take the bees away, or make them too scarce and too unreliable to count on, and the brilliant machine produces exactly nothing. A perfect outcrossing trap in an empty world is a flower that never sets a seed. The con only works on a crowd.

Go far enough north and the crowd thins out. Up in the boreal forests and the sub-Arctic, on ground the glaciers scraped bare and then released less than ten thousand years ago, the season is short and cold and the pollinating insects are few and chancy. And there, on that young raw ground, lives a small white-pouched lady's slipper that gave up the game entirely.

It is the sparrow's-egg lady's slipper (*Cypripedium passerinum*), its small white pouch washed with a faint purple freckling that shows through the translucent walls. Open one of its flowers and the trap is all still there, the inflated pouch, the inrolled rim, every part it inherited from the same deceiving ancestor as the rest. But the sparrow's-egg slipper no longer uses it. It pollinates itself, with no insect at all, by a small mechanical surrender that took botanists by surprise when they first described it. Biologists call self-pollination **autogamy**, the flower fertilizing itself with its own pollen, and the sparrow's-egg slipper is the first lady's slipper in which anyone ever documented it, reported from the north shore of Lake Superior in 1983 and worked out more fully in the years after. The mechanism is almost demure. The little stalk that holds the stigma is shortened, so the stigma sits right up against the flower's own anther. And the pollen, which starts out sticky the way a lady's slipper's pollen should, turns over a few days after the flower opens from sticky to nearly fluid, and simply slides down onto the waiting stigma under its own weight. No bee. No trap sprung. No sieve to pass. Gravity does the work the queen bumblebee does in the pink slipper and the leaf-cutter does in the showy one. The flower marries itself.

And the reason is the cold young ground it lives on. On terrain the ice left bare within the last ten thousand years, where a pioneer plant is colonizing fresh country and the pollinators are sparse and the summer is brief, a flawless outcrossing machine that depends on a fooled bee is a liability, not an asset. It may set no seed at all in a season when the right insect never comes. So the sparrow's-egg slipper made the opposite bet from every cousin in this chapter. It traded the genetic richness of outcrossing for the plain certainty of a seed. Selfing guarantees a full pod even when no bee shows up, and on the frontier of fresh ground, a guaranteed seed beats a brilliant gamble. Where the others stayed clever, this one chose to be sure. (Not every native lady's slipper that abandoned the bee turned to selfing, either; one small dark western species hands the job to a parasitoid wasp hunting fungus-gnat hosts, but that is a mountain story, and I will keep it for the mountains.)

I want to be honest about the sparrow's-egg slipper, though, because the truth has one more fold in it and the fold is the best part. It did not abandon the trap completely or everywhere. Across most of its northern range it selfs, predominantly and reliably; but in one place it still does the old thing. In the far northwest, in a corner of the southwest Yukon that the glaciers somehow never reached, an ancient unglaciated refuge where the land was never scraped bare and never had to be recolonized, botanists have found sparrow's-egg slippers built the other way, with the anther held apart from the stigma so the pollen cannot fall on it, flowers that still need an outside vector, that still keep faith with the bee. The one place the orchid never had to flee and refound itself is the one place it never gave up outcrossing. The exception has its own exception, and the exception's exception is a window straight back into what the whole species used to be. We do not fully understand it yet. That is not a flaw in the story. It is the live edge of it, the part still being read.

The bee gets free

Our sweat bee is still in the yellow pouch where we left her, and it is time to let her finish.

She has done what trapped things do. She climbed toward the brightest part of that dim chamber, toward the two lit windows at the back, up the slope of stiff hairs that all lean the same way, and she has found the only gap that gives. It is a tight one. She wedges herself into it and pushes, scraping along the narrow passage beside the column, brushing the sticky stigma as she goes and then, a fraction of a second later, the anther, which smears its pollen onto her back where she cannot reach to clean it off. It is not the neat waxy pellet most orchids hand their pollinators. A lady's slipper's pollen is mealy and tacky, the way I told you back in the orientation chapter, and it goes on her in a loose sticky daub rather than a tidy packet. She works her shoulders through. For a moment she is half in and half out, legs scrabbling. Then she is free, out the back of the flower into the open air, dusted, cheated, no heavier with food than when she went in.

She rests a second on the lip of the pouch. Then she lifts off, the new yellow grains riding on her back, and flies low across the clearing, looking for the next promising flower, carrying this one's future to wherever she lands. If she is fooled again by another yellow slipper, and not by the same plant she just left, the orchid will have won its slow rare bet for one more year.

I stay crouched where I am a while after she's gone, looking at the pouch she came out of, still nodding a little on its stem in air I cannot feel move. From the outside it is only a yellow flower again. You would never know, to look at it, that it is a machine for choosing bees, or that a smaller cousin of it far to the north long ago grew tired of choosing and learned to need no one at all.

Part Three — The Cast

There are twelve native lady's slippers in North America, and no two of them solve the problem of being a lady's slipper the same way. Here they are, one at a time: the pink and the showy, the prairie white and the ram's-head, the serpentine specialist of the western seeps and the palomino of the Aleutian fog.

The Pink One

The first time I saw pink lady's slippers growing like a weed, I was a long way from home, in uniform, on the Lake Huron shore.

The Coast Guard had stationed me there, and somebody told me to go see the orchids in the pine woods on the old dune ridges up the coast. I went expecting the thing I knew. Back home in West Virginia the pink lady's slipper is a once-a-summer piece of luck, a flower I had spent years learning to find one at a time on the acid ridges, kneeling, parting the leaf litter, going quiet when I succeeded. What I found instead was pink pouches by the hundred, nodding in the cool air off the lake, thick under the pines the way buttercups grow thick in a pasture. There is a town along the shore that holds a festival for them every Memorial Day weekend, timed to the bloom. A festival, for the flower whose every spot I had guarded like a secret.

It is not only Michigan. Off in the Maritimes the pink lady's slipper is so common it is the provincial flower of Prince Edward Island, printed on the license plates, sold on tea towels, thick in the roadside ditches. And yet across most of its enormous range it is exactly what it is for me at home: the rarest thing most people will ever stumble on, a plant whose haunts you keep quiet. Same flower. The contrast told me something true about what this orchid actually is, and it is the opposite of what nearly everyone, including me, had assumed.

This is the pink lady's slipper (*Cypripedium acaule*). It is the flower this whole book opened on, and the one most of us half-recognize without ever having truly looked at. It is time we met it properly.

Pink lady's slippers by the hundred in open pine woods near the Lake Huron shore. The same species can be a roadside crowd in one place and a rare find in another.

The one you already half-know

Of the twelve native lady's slippers in North America, this is the famous one. If you have ever seen a wild orchid in the eastern half of the continent, the odds are heavy that it was this. If a stamp or a state emblem or a faded nature-center poster has lodged a "lady's slipper" somewhere in your memory, it is very likely this pink pouch you are remembering. And yet it is also, in a way I find almost funny, one of the least examined plants of its fame. The most celebrated native orchid we have is the one whose scent has never been run through a laboratory, whose fungus has been sampled in one corner of one state, whose pollination Charles Darwin himself gave up on. Famous and unread at the same time. We will get to all of that.

Start with the body, because two features tell you it is this orchid and not one of its eleven cousins, and you can learn both in a single look.

The first is at your feet. Where every other native lady's slipper carries its leaves up a leafy stem, *C. acaule* has none up there at all. The flower rises on a single bare stalk, a **scape**, naked the whole way up, and the leaves stay below, two of them, opposite each other, pressed close to the ground and pleated like a folding fan. Two basal leaves and a leafless stalk. No other native *Cypripedium* on the continent does this; every one of the others runs its leaves up an aboveground stem. The Latin says so plainly. *Acaule* means "stemless," from the Greek for "without a stalk," and as orchid names go it is unusually honest about what it describes. I noted in the first chapter that this is one of the few that tells the truth; here is the truth it tells. The plant put its leaves at the bottom and sent up nothing but a flower on a pole.

The second feature is the pouch itself. I walked you through it slowly once already, so I will only point: every other lady's slipper opens its inflated pouch through a round hole near the top, like the mouth of a small jug, and this one opens down the front, through a long vertical slit with its two edges rolled inward on themselves. That fissure is found on no other *Cypripedium* anywhere on Earth. It is the orchid's signature and, as we saw, its trap. You do not need to re-run the con to know the flower. You need only those two facts: leaves at the ground, slit down the front. After that you will never misidentify it again.

The slit down the front of the pink lady's slipper pouch, its two edges rolled inward: the one-way door found on no other Cypripedium.

Now stand it up and let yourself see the size of it. It is a small thing. Knee-high at most. The bare scape rises somewhere between fifteen and sixty-one centimeters (six to twenty-four inches), though most of the ones you meet reach you about mid-shin. And the pouch itself, the whole reason you stopped, is smaller than people remember: thirty to sixty-seven millimeters (about one and a half to two and a half inches), about the size of the top joint of your thumb. The color runs from a soft clear pink through magenta to a deep rose, often with darker veins threading the front. Here and there you will also find the ghost of it — a rare pale form, very nearly white, the pink drained almost entirely out of the pouch. Botanists call it forma *albiflorum*, and to come on one in a wood full of the ordinary pink is a small electric event.

Now picture where that small thing lives, because the range is enormous and worth feeling as geography rather than reading as a list. Start in the southern Appalachians, in Georgia and Alabama, and walk north. The pink one comes with you the whole way: up the spine of the mountains through the Carolinas and Virginia, out across the Pennsylvania ridges and the New Jersey pine barrens, into New England and up to Maine and the Maritimes, the cool granite edge of the continent. Swing west and it reaches the Great Lakes and Minnesota, then climbs north into the boreal forest of Canada, province after province, until the only Canadian province without it is British Columbia, off on the far side of the Rockies where the pink one never arrived. That is a single orchid's territory. Most of the eastern half of a continent.

And it does not bloom all at once. The flowering moves like a slow wave up that map. In the southern Appalachians the pink pouches open in late April, while the northern ground is still cold and bare, and the wave rolls north through May and does not reach northern Maine and the Maritimes until June or July. The Lake Huron woods where I saw them by the hundred bloom toward the end of May, around Memorial Day; the roadside ditches of Prince Edward Island, farther north and east, do not break until June, a thousand miles behind the first Georgia flowers of the same spring.

One more thing to hold before we go underground, and it makes everything in this chapter weigh more. A pink lady's slipper does not flower quickly. From the dust seed I described in an earlier chapter to the first pink pouch on a scape takes, in the wild, something like six to ten years, almost all of it underground and unseen, fed by the fungus, building toward a bloom no one is watching for. The flower you stopped for is the visible tip of the better part of a decade. Keep that number; we will need it at the end of this chapter, when somebody steals thirty of them out of the ground.

So that is the pink one to look at, and to walk a continent finding. But the contrast between those crowded Michigan woods and the lean ridges back home left me with a harder question than what the plant looks like. Where does it actually live, and why there?

The plant that took the sour ground

Here is a thing the field guides tend to bury in a clause: this orchid is an acid specialist, and a severe one.

Most of the showy lady's slippers, the ones people drive miles to see, are creatures of sweet ground. They want lime in the soil, the chalky residue of old seabeds or limestone, the kind of mineral-rich wet meadow we call a fen. The pink one wants the opposite, and wants it badly. It grows where the soil has gone genuinely sour, down around pH four to four and a half, well below the comfortable middle of the scale where most plants and most soil life do their living. Think of the places that go that acid. The duff under a stand of white pine and hemlock, where a century of fallen needles has steeped the ground like strong tea. The thin droughty soils of an Appalachian oak-heath ridge. The sand of the New Jersey Pine Barrens. The hummocky margin of a black-spruce bog, in among the leatherleaf and the sheep laurel and the cranberries. The pink lady's slipper lives in all of them. The ground is poor. Starved and sour enough that most plants find it not worth the trouble.

We have a habit of reading a plant like this as fragile. A delicate thing, we say, clinging to existence in a hostile place, one bad year from being lost. I held some version of that picture myself for a long time. It is wrong, and those crowded Michigan woods are the proof. This is not an orchid barely surviving the sour ground. This is an orchid that took the sour ground, the starved, acid, picked-over ground that every richer plant declined, the ground nobody else wanted, and made it home, and owns it. Not fragile. Particular. There is a difference, and it is the whole character of the plant.

I told you back in the chapter on the fungus why the acid matters, so I will only touch it here and move on. The sourness is not an obstacle the orchid endures. It is a wall the orchid hides behind. In sweeter soil the riot of competing microbial life would swamp the one narrow fungal partnership the lady's slipper cannot live without; the acid holds that riot down and keeps the partner safe. The pink one's whole life is arranged around keeping faith with something underground. What I want to add here is what that arrangement *costs*, and the strange shape of the threat it creates.

Because the pink lady's slipper is not merely fond of acid soil. It is trapped in it, in both directions. Push the ground much above pH five and the orchid sickens; its roots rot, the partnership fails, the plant dies in soil most flowers would call ideal. This is the deepest reason a dug-up wild plant almost never lives: the severed plant coasts a year on its rhizome, then starves when the fungus it cannot carry fails to take hold in your too-sweet garden bed. The pink one is the worst case in the whole genus, with a transplant survival rate of about five percent. You cannot move it because you cannot move the sour ground and the buried partner together, and it will accept no substitute for either. The plant is faithful to a fault. The fault is what kills the ones we steal.

*The pink lady's slipper (*Cypripedium acaule*), its deep-pink pouch veined and swollen on a single nodding stalk. Not fragile. Particular.*

That faithfulness to acid soil gives the pink one a conservation worry no other lady's slipper shares, and it is a genuinely strange one.

For most of the twentieth century, much of eastern North America rained acid. The sulfur our power plants and tailpipes put into the sky came back down in the rain and slowly soured the soils, decade after decade, especially across the Northeast and the Appalachians, the very heart of this orchid's range. We noticed, eventually. We passed laws. We scrubbed the sulfur out, and the rain grew cleaner, and the soils began, very slowly, to recover, to climb back from the acid trough of the nineteen-eighties part of the way back toward the sweeter chemistry they held before all that industry. This is, by every ordinary measure, good news. Cleaner rain, healthier streams, forests no longer steeping in acid.

But picture what slow recovery means for a plant that needs the soil sour. As the ground sweetens back toward neutral, the very acidity the pink lady's slipper lives by may ebb away beneath it, the shelter thinning not torn down but simply, gradually, eroded by our success at cleaning the air. It is the one orchid in this book for which a healing landscape might be a quiet kind of threat. No one has yet watched this happen to *C. acaule* and measured it. The soil chemistry is documented; the consequence for this particular plant is a reasoned worry, not a proven decline. Still, sit with the shape of it. Nearly everything we do to a wild plant, we do by harming its world. Here is one whose world we may be subtly unsettling by *mending* it. The pink one would, if it could weigh in, ask us to leave a little sourness in the ground for it. There is something humbling in a plant whose ideal is the thing we worked so hard to undo.

The moccasin flower

The plant has older names than the Latin, and the oldest of them is the one most Americans still half-remember: the moccasin flower.

The name is not a poetic flourish dropped on the orchid by a settler with a thesaurus. It is older than that, and it belongs to a people, and the careful thing is to say so exactly. Across much of this orchid's northern range lie the homelands of the Anishinaabe, and in Anishinaabemowin, their language, the flower carries names built from the word *makizin*, moccasin, shoe. The Great Lakes Indian Fish and Wildlife Commission records the name *makizinkewe*, "moccasin-making one," in its 2007 guide *Indinawemaaganidog*, which translates as *All My Relatives*. The names are written down. The root *makizin* and the fuller form *makizini-waabigwan*, "moccasin flower," are documented in the Ojibwe People's Dictionary at the University of Minnesota, a project whose entries carry the recorded voices of Anishinaabe speakers. And the most substantial contemporary trace I know of is a December 2024 piece by the Ojibwe scholar Anton Treuer, of Bemidji State University, titled *Makizini-waabigwan: Anishinaabe Aadizookaan*, the moccasin flower as the subject of a traditional story, told in Ojibwe.

Beaded hide moccasins, attributed to an Anishinaabe (Ojibwe) maker, about 1890 to 1910. The flower's oldest English name, moccasin flower, translates the Anishinaabemowin name for the shoe. Minneapolis Institute of Art, public domain.

I raise that last point with care, because it is the most sensitive thing in this book. There is an Anishinaabe story tied to this flower. It belongs to the Anishinaabe. It has been published in several forms and is carried still by Anishinaabe storytellers, and from the outside one can say only the barest outline of it: that it tells of a girl and a great winter need and the small footwear that gives the flower its name. That is as far as I will go, and even that I offer cautiously. I am not the one to tell the story, and I am not going to retell it here. I am a settler and a learner, sitting at the edge of a tradition that is not mine, and the most respectful thing a person in my position can do is to point and say: this exists, it is older than any field guide, it is held in trust by the people whose story it is, and if you want it, go to them. The English name "moccasin flower" is a worn-down translation of a far richer relationship. That relationship came first. The botanists arrived centuries later, and so did the word "lady's slipper," which only displaced the older name among most

Americans in the nineteenth century. When you say *moccasin flower*, you are speaking, however faintly, in a borrowed tongue, and it is worth knowing whose.

What the settlers made of it

The Anishinaabe knew this flower as a relative with a story. When nineteenth-century American and English writers got hold of the moccasin flower, they made something quite different of it, and the difference is worth looking at squarely.

Some of what they made was simple field affection. John Burroughs, the great popular naturalist of the eighteen-nineties, knew the pink one well and noticed exactly the thing this chapter began with: that it keeps to dry, sour ground while its cousins prefer the wet. He wrote, in his 1894 book *Riverby*, that "most of the floral ladies leave their slippers in swampy places in the woods; only the stemless one (*acaule*) leaves hers on dry ground." I love that he tucked the orchid's real habitat into a joke about footwear. The English botanical illustrators had been less kind. When *Curtis's Botanical Magazine* figured the plant in 1793, the text sniffed that "the present plant in this respect is far less handsome than *C. Calceolus*" (the European yellow). It says more about the homesickness of London botanists for their own slipper than about the pink one nodding in an American pine wood. Admired and condescended to in roughly equal measure since the day it entered the books.

The settlers made a third thing of the moccasin flower, though, and this one needs naming for what it is. They turned the Indigenous name into an emblem of Indigenous death.

In the eighteen-hundreds a strain of American and English Romantic poetry took the "moccasin flower," a name carrying an Indigenous relationship the poets did not share and did not honor, and bent it into a symbol of the vanishing Indian, the noble native fading from a landscape the writers were in the very act of taking. Elaine Goodale, a young New England poet, published a poem simply called "Moccasin Flower" in 1879, and at its gentlest it is a pretty woodland reverie. She calls the orchid "shy and proud among the forest flowers, / In maiden solitude," an image of demure wild beauty that asks nothing uncomfortable of the reader. I quote that line and stop there on purpose, because elsewhere the same poem reaches for the freighted equation of the flower with the displaced Indian, and that part of it is not mine to extend or to dress up as nature writing. A generation earlier William Cullen Bryant, the most eminent American poet of his day, wrote a poem called "The Maiden's Sorrow" set at a grave in "the distant west," and let the flower mark the burial: "There, in the summer breezes, wave / Crimson phlox and moccasin flower." The moccasin flower, in Bryant's hands, grows on a grave. The Indigenous name has become a headstone.

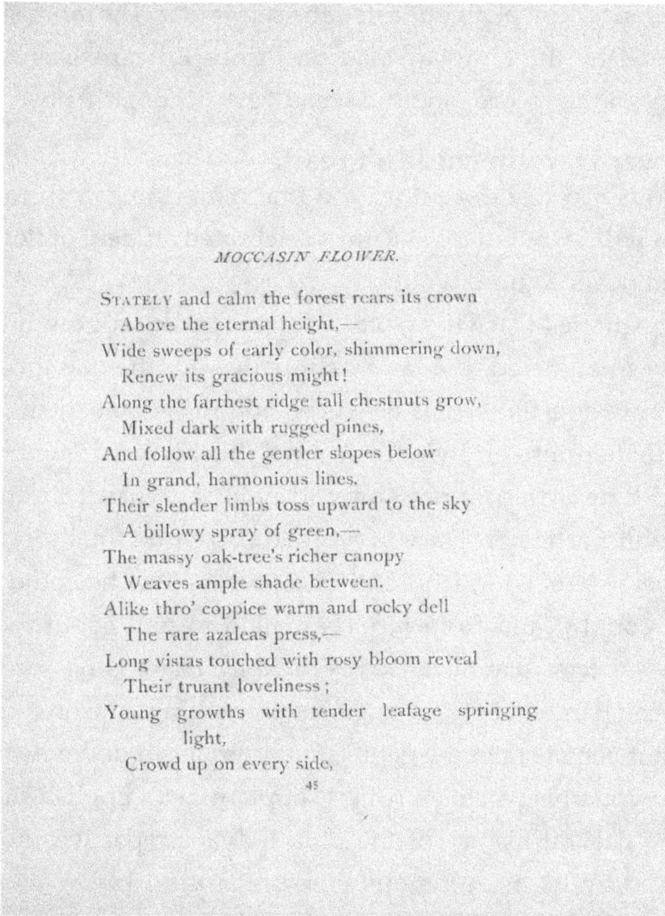

MOCCASIN FLOWER.

Stately and calm the forest rears its crown
  Above the eternal height,—
Wide sweeps of early color, shimmering down,
  Renew its gracious might!
Along the farthest ridge tall chestnuts grow,
  Mixed dark with rugged pines,
And follow all the gentler slopes below
  In grand, harmonious lines.
Their slender limbs toss upward to the sky
  A billowy spray of green,—
The massy oak-tree's richer canopy
  Weaves ample shade between.
Alike thro' coppice warm and rocky dell
  The rare azaleas press,—
Long vistas touched with rosy bloom reveal
  Their truant loveliness;
Young growths with tender leafage springing
    light,
  Crowd up on every side,

45

The opening of Elaine Goodale's 1879 poem "Moccasin Flower," settler verse that bent an Indigenous name toward the "vanishing Indian" trope. It is the artifact this section reads critically, not admired verse. Public domain.

I do not quote these to honor them, and I will not pretend they are neutral pictures of a plant. They are part of this orchid's documented record in settler literature, and an honest book acknowledges that record and names it. What Goodale and Bryant were doing was a move the period made constantly and we should be able to see clearly now: taking something Indigenous, in this case a name and a relationship, and repurposing it as decoration for a sorrow the writers themselves were helping to cause. The flower the Anishinaabe knew as a relative with a story became, in the settler poem, a tasteful emblem of Native disappearance. The plant outlasted the sentiment. The moccasin flower is still here, by the highway-load on Prince Edward Island. But the poems are in the record, and they read very differently now.

The flower Darwin couldn't read

The famous nineteenth century did one more thing with this orchid, and what it did was fail to. The most celebrated student of flowers who ever lived never figured it out.

When Charles Darwin wrote his great book on how orchids get themselves pollinated, the lady's slippers were the one group he set aside, confessing flatly that he knew nothing about them. We have crossed that admission twice already in this book, so I will not quote it a third time; I raise it now only to point at what it means for the pink one in particular. Darwin worked, when he finally returned to the slippers years later, from the European yellow he could watch in his own country and from reports mailed to him about a couple of American species. He never saw *C. acaule*. There is no record of his ever having handled it, no note, no letter, nothing. The orchid with the strangest mechanism in the genus, that unique slit down its front, was simply beyond his reach: an American plant in an English study.

So the natural history of the pink lady's slipper was not written in England by its most famous possible author. It was written here, slowly, by Americans, over more than a century after Darwin set the genus down. Botanists working in Virginia and New Brunswick and across the eastern woods were the ones who watched the queen bumblebee force the slit, who learned that the books had blamed the wrong, smaller bees for years, who counted the dismal fruit set and worked out that this gorgeous flower is pollinated only rarely and only by the strong. It is easy to imagine that all the important looking was done long ago by famous men in Europe. It was not. The most-recognized native orchid on the continent got its biography in American pine woods, from patient local observers whose names you will never see on a stamp — watching a flower the most famous naturalist alive had given up on.

The emblem, and the orchid it replaced

Which brings us back to the island, and to the strangest piece of this orchid's cultural life: the places that have made it a flag.

In 1991, New Hampshire named the pink lady's slipper its official state wildflower, a distinction separate from the state flower (which remains, oddly, the purple lilac, a plant from southeastern Europe). And in 1965, Prince Edward Island made it the provincial flower; the pink pouches fill its roadside ditches every June.

But the island's choice carries a small drama that the New Hampshire one does not, and it is the perfect closing fact for this orchid's character. Prince Edward Island had a floral emblem before 1965. It was the showy lady's slipper, *C. reginae,* the tall white-and-pink queen of the genus, the province's earlier emblem. The trouble was that the showy lady's slipper, on that island, had grown too rare to serve. The very thing that makes the showy one magnificent, its fussiness, its appetite for specific wet calcareous ground, had thinned it on Prince Edward Island until it was hard to find. So in 1965 the island did something I know of nowhere else in North America. It demoted one lady's slipper and promoted another. It let the rare, regal showy one go, and raised up the pink one in its place: the common one, the roadside one, the one that grows by the hundred where the soil is sour, the one no one has to drive far to find.

I love this for what it quietly says. The island did not choose its emblem for rarity or grandeur. It chose the flower its people could actually walk out and meet, abundant enough to belong to everyone. The showy lady's slipper is the more spectacular plant, and it gets its own chapter later. But there is a deep rightness in a place choosing, for its symbol, not the queen it could rarely find but the patient pink commoner it could not avoid finding.

A secure flower that keeps disappearing

I should not leave you thinking the pink one is safe everywhere just because it is safe on an island. The full reckoning of what we owe these plants belongs to the end of this book, but the pink one carries one lesson about conservation so cleanly it would be a waste not to set it down here.

On the standard global rankings, *C. acaule* is rated secure, the most reassuring category there is, the one given to plants in no danger of vanishing from the world. And on the whole that rating is honest. There is no real prospect of losing the species; it is too widespread, too abundant in too many places, the pink tide on the island and the great washes of it along certain northern highways. But "secure" is a verdict about the whole range at once, and a range is not a neighborhood. The same orchid that runs riot on Prince Edward Island is listed as endangered in Illinois, where it may already be gone, and clings to the northern edge of Indiana, and is protected by statute in Tennessee against commercial collection, and sits on the endangered lists of Alberta and Manitoba out at its western boreal limit. A plant can be globally secure and locally vanishing at the same time. The reassuring number at the top hides a hundred small losses underneath.

And the losses are not abstract. In June of 2020, somebody walked onto federal land at the Big South Fork, on the Tennessee-Kentucky line, and dug up about thirty pink lady's slippers. Stole them out of public ground, presumably to sell or to plant, almost certainly to kill, given the five-percent odds. Thirty plants, each of them the visible tip of a buried life that took the better part of a decade to reach its first flower and might have stood there for many decades more. The thief did not steal thirty flowers. The thief stole thirty long slow histories out of the ground, and almost none of them will have survived the move. That is what "locally vanishing" looks like up close. It looks like a hole in a hillside where the pink used to be.

Back on the ridge

Years after Michigan, I still find them at home the way I always have: one at a time, on the dry acid ridges, where the pine duff lies thick and sour underfoot and you can walk a whole slope to turn up a single pouch. I do not dig them. I have never needed to; that is rather the whole point of a plant like this, that on the right ground in the right week it asks nothing of you but to be looked at.

A bumblebee forcing her way in through the front slit. Only a bee this strong can open the door; a smaller one cannot. The con in action.

I think of the Michigan woods sometimes when I am up there, the hundreds of them in the cool air off the lake, and the one pouch at my feet does not look to me like a poorer version of that crowd. It is the same flower doing the same thing it does everywhere: making a home of the sour, starved ground that defeats almost everything else, by the hundred where the ground allows and one at a time where it does not. Not fragile, here on the lean slope. Particular, and entirely at home. A single pink pouch, slit down the front, rising on a bare stalk from between two pleated leaves pressed to the sour West Virginia ground, nodding in air too still to feel: somewhere down in that duff, finer than thread, the fungus is holding the coils and passing the carbon, keeping faith the way it always has. A bumblebee blunders up the slope, in and out of the slit, fooled and fooled again. And the flower the most famous naturalist alive could never get his hands on, the flower a whole century of poets turned into a headstone, the flower that took the ground nobody wanted, goes on doing exactly what it does, on the ridge, in the quiet, where anyone who slows down can find it.

The Yellows

There is a yellow lady's slipper blooming in a limestone wood in late May, and its name is a lie.

The pouch is big. Held up against a beam of light coming through the canopy, it is the size of a small hen's egg, the color of fresh butter, the front of it freckled with tiny red dots and threaded with darker veins, and off to either side two long petals twist away from it like ribbon that someone has run between thumb and fingernail. It is one of the largest, showiest wildflowers in this whole eastern wood. And the species it belongs to is called *Cypripedium parviflorum*, the small-flowered lady's slipper. *Parvus*, small. *Florum*, of flowers. The botanist who named it in 1791 called the most magnificent of its forms small, and the name stuck, and it has been wrong about this particular pouch for more than two hundred years.

That is the small joke. Here is the large one. For roughly a hundred and fifty years after that, an even bigger lady's slipper was hiding among the big yellows of the American South, the largest-flowered slipper orchid in the entire world, and nobody recognized it as anything but an oversized one of these. Botanists alive today went and found it. They gave it a name in 1981, within the lifetime of people you could call up on the phone. The biggest slipper orchid on Earth spent a century and a half mistaken for a yellow lady's slipper that had simply grown a little large.

This chapter is about a flower that botany is, in a real and ongoing way, still sorting out: the familiar yellow one, beloved and everywhere, and the hidden giant of the South that hid inside it for a century and a half. The line between them turns out to be exactly the kind of line that is hard to draw, and the reason it is hard to draw is the whole story.

*A yellow lady's slipper (*Cypripedium parviflorum*), photographed wild in Minnesota: the waxy butter-yellow pouch and the long, spirally twisted side petals that mark it.*

The familiar one

Start with the yellow lady's slipper (*Cypripedium parviflorum*), because if you have walked the woods of eastern North America in spring with your eyes open, you have very likely met it, and because it is, by a wide margin, the most widely distributed lady's slipper on the continent.

It grows almost everywhere. North to Alaska and the Yukon, south to New Mexico and Arizona, east to Newfoundland, west to the Rockies, present in the great majority of states and provinces, a reach no other native slipper comes close to. Where the pink one we met in the last chapter is a creature of sour, acid ground, the yellow one is mostly the opposite: it favors sweet ground, soil with lime in it, limestone woods and dolomite glades and the rich black muck of a calcareous fen. One orchid took the acid. This one took the lime. You can read a good deal of where you are standing from which lady's slipper greets you there.

Now look at the flower itself, because two things mark it.

The first is the color. The pouch is yellow, sometimes a pale lemon, sometimes a deep gold, and inside the front it is dotted and streaked with red or purple-brown, as if someone had flicked a wet brush at it. The second, and the one I want you to carry, is those petals. A lady's slipper has two lateral petals flanking the pouch, and in this species they are long, narrow, and **spirally twisted**, corkscrewing away to each side like a ribbon pulled taut and curled. No other eastern lady's slipper does it quite so plainly. Once you have seen those twisted petals you can name the yellow one across a clearing, before you are close enough to count anything else. Yellow pouch, twisted ribbons. That is the field mark.

The lateral petals of the yellow lady's slipper, corkscrewing away on either side of the pouch: a dependable field mark.

And it is a leafy plant, which matters for telling it from the pink one. Where *C. acaule* keeps its two leaves flat on the ground and sends up a bare pole, *C. parviflorum* runs its leaves up the stem, three to five of them, pleated and clasping the stalk, with a single flower (now and then two) nodding at the top. Stems run from about knee-high to nearly waist-high in the largest plants, twelve to eighty centimeters, say five to thirty-two inches.

A cultivated yellow lady's slipper (Cypripedium parviflorum)*, several pouches to a clump, photographed by the author's mother, Kim Zyla-Scott, from her garden.*

So far, so straightforward: a leafy, lime-loving, twist-petaled yellow orchid across most of the continent. But that tidy description hides what the name *parviflorum* hides. There is no single yellow lady's slipper. There are several, they shade into one another, and the business of sorting them out is where this chapter's real subject begins.

The varieties, and the name that lies

Botanists currently recognize four varieties of the yellow lady's slipper, and the cleanest way to feel the problem is to walk through them by size, because size is where the famous name comes undone.

Line them up. There is the small eastern one, variety *parviflorum*, the original, the plant the 1791 name was actually built on, with a modest pouch an inch or so long and densely red-dotted petals. There is variety *makasin*, the northern fen plant, small-pouched too, tucked into cedar swamps and alkaline wetlands across the boreal band, with an intensely sweet, vanilla scent. There is the far-northern variety *exiliens*, slender and small, described only in 2010 from the Brooks Range of Alaska and the Yukon, the most recently named of the four. And then there is variety *pubescens*, the big one, the great butter-yellow pouch the chapter opened on, three to more than five centimeters long (about one and a quarter to two-plus inches), the plant that stops you on a trail.

Now hold the name up against the lineup. *Parviflorum* means small-flowered. Three of the four varieties are, honestly, small-flowered — the eastern one, the fen one, the slender northern one. And one of them, the most familiar and conspicuous of the lot, the one most people picture when they hear "yellow lady's slipper," is anything but small. The species name describes three-quarters of itself and flatly contradicts the rest, a piece of botanical bookkeeping that fell out of priority: the small one was named first, so when the big pouch was folded into the same species a century later, it had to live under a label that calls it tiny.

I could walk you through the rest of the distinctions, the pubescence of a floral bract, the precise shade and spotting of the petals, the four different scents. But I want to be careful not to oversell how sharp those lines are, because the botanist who did more than anyone to define these four varieties did not oversell them himself. Charles Sheviak, who spent a career on this genus and named variety *exiliens*, called the characters that separate the varieties "minor and inconstant." That is a striking thing for the describing authority to say. The man drawing the lines admitted, in print, that they wobble. The yellow lady's slipper does not come in four neat boxes. It comes in a smear. Botanists have drawn four boxes around that smear, knowing the boxes leak.

Which raises the obvious question. If the varieties are so hard to tell apart, so "minor and inconstant," why are they varieties at all? Why not call them one plant and be done?

The answer is the heart of this chapter, and it took modern genetics to give it to us.

A flower caught mid-sentence

In 2024, a team of botanists led by Loudmila Lagou, working in Munich, did something nobody had been able to do before with this orchid. They read its genome at scale, more than nine hundred separate stretches of nuclear DNA, across the yellow lady's slipper and its relatives, and they built the family tree from the inside out, letter by letter.

What they found explains the whole muddle.

The yellow lady's slipper, read genetically, does not behave like a finished species. A finished species is a closed circle: every member shares a common ancestor that no outsider shares, and the circle holds. The yellow lady's slipper does not hold. The technical word the geneticists use is **paraphyletic** — the group, as we have defined it, leaves some of its own descendants outside the circle and folds an outsider in. Its varieties do not branch cleanly off one trunk. They tangle. And the cause, the team concluded, is two things happening at once: lineages that have not yet finished separating, and lineages that separated partway and then leaked back into one another, exchanging genes across the gap.

There is a name for that second process, and it is worth setting down because it is the engine of this whole story. Biologists call it **reticulate evolution** — evolution that does not run as a tidy branching tree but as a net, with the branches reaching back across the gaps between them and fusing, so that the lineages braid instead of merely splitting. The yellow lady's slipper is not a species that has finished dividing into four. It is a population in the middle of dividing, the varieties best understood not as four finished things but as **incipient species**: lineages in the act of becoming separate, caught partway, still trading genes across the fault lines that are slowly opening between them. Beginning species. Not yet done.

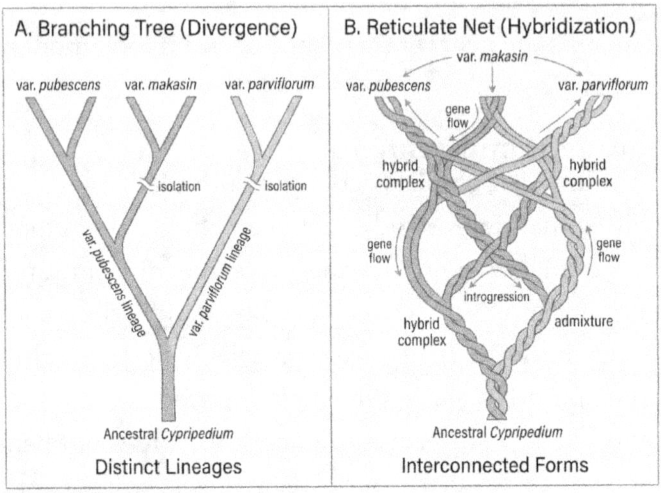

Not a tidy family tree but a braided net. The yellow lady's slippers are still trading genes across their varieties, lineages fusing as much as splitting.

Sit with that, because it reframes everything, including your own confusion in the woods. When you stand over a yellow lady's slipper and cannot quite decide which variety you are looking at, you are not failing at botany. You are seeing the truth of the plant. It is hard to tell these forms apart because *they are not yet fully separate things*. Botany cannot draw a clean line through them because the line does not yet exist. We have caught it mid-sentence, and the sentence is not finished.

And here is the hinge the rest of the chapter swings on. When Lagou's team read the tree, the big ivory southerner we are about to meet, the largest slipper orchid in the world, did not sit off on its own clean branch either. It came out so tangled against the yellows, so close to them and so recently split from them, that botany still cannot say exactly where on the tree it sits. Which is exactly why, for a hundred and fifty years, nobody could see it was its own thing. It was hiding inside an unfinished sentence.

The fragrant ones

Before we go south, there is one more thing about the yellows that I cannot leave out, because it is the sense most field guides forget you have. The yellow lady's slippers are the fragrant ones, and the varieties do not all smell the same.

Crouch to a flower of the fen variety, *makasin*, on a warm morning, and the scent is sweet and clean, something close to vanilla. The big *pubescens* smells different, warmer and rounder; people reach for words like honey and apricot. These are not my impressions alone. They are field marks. The difference is not only in the nose, either. In 1992 a team of Swedish chemists led by Gunnar Bergström did the only modern laboratory analysis of scent that exists for any North American lady's slipper. They captured the volatile compounds coming off different yellow forms and ran them through their instruments, and they found the scents dominated by chemically different classes of compounds, variety to variety. The plants are not just faintly different to a human nose. They are built on distinct chemical recipes, which the chemists read as each form tuning its advertisement to the bees of its own region. The exact mapping of which class of compound belongs to which variety has a wrinkle in the record I do not want to overstate, so I will leave it where it honestly sits: the varieties smell distinctly different, and a careful 1992 analysis tied those differences to different classes of volatile compounds.

Even the difference in scent, you will notice, is one more way the varieties are not quite the same and not quite different: distinct enough to smell, blurry enough to argue about. The pattern holds all the way down to the molecules.

This is the kind of flower that gets noticed, and it has been getting noticed for a long time. The naturalist Neltje Blanchan, writing about the yellow lady's slipper in 1900, in a frankly enraptured turn-of-the-century voice, said this orchid, "swinging outward from a leaf-clasped stem," "attracts us by its flaunted beauty and decorative form from tip to root." *Flaunted.* It is exactly the right word for a plant that is loud about itself, that smells of vanilla and waves its twisted ribbons and dares you not to stop. Blanchan saw a show-off. She was not wrong.

The names, and whose names they are

The yellow lady's slipper carried names long before any botanist reached for Latin, and two of them belong to peoples whose relationship with this plant runs centuries deeper than the field guide, and the careful thing is to say so exactly.

In 1891 the ethnographer James Mooney recorded, among the Cherokee, a name for the yellow lady's slipper that I find quietly perfect: *gŭgwĕ'-ulasu'la,* which translates as "partridge moccasin." A small shoe, a bird's shoe: the same seeing-a-slipper-in-the-flower that runs through so many of this plant's names, but tuned to a partridge rather than a person. I set the name down because it is documented and because it is older than the botany, and I leave it there, pointing rather than explaining. It belongs to the Cherokee, and I am a settler and a learner standing at the edge of a tradition that is not mine to interpret.

The second name is stranger, because it did not stay in the Indigenous language. It crossed over and lodged itself inside formal botanical Latin, where it sits to this day. Remember variety *makasin,* the sweet-scented fen plant. That word, *makasin,* is the Anishinaabemowin word *makizin,* moccasin, shoe, the same root the whole continent borrowed for the footwear. When the botanist Oliver Farwell named that variety in 1917, he did not translate the Indigenous word or replace it with a Latin one. He spelled the Anishinaabe word into the scientific name and left it there. So in the official, Latinized, internationally regulated name of this plant, *Cypripedium parviflorum* var. *makasin,* an Anishinaabe word rides along, preserved, a piece of one language carried inside the formal name-system of another. Every botanist who writes that variety down writes down a moccasin in Anishinaabemowin, whether they know it or not.

And there is a harder strand braided in with the names, which is what the settlers did with the plant once they learned of it. Through the nineteenth century the yellow lady's slipper was dug from the ground by the ton and sold as a medicine called "American Valerian," or nerve root, a sedative listed in the official United States Pharmacopoeia for decades. I will save the full weight of that trade for a later chapter, where it belongs, because it is mostly a conservation story and a long one. But it has to be named here, because the medicinal knowledge the trade was built on was not the settlers' to begin with. The plant was a medicine in Indigenous hands first, across many Nations, long before a single pound of root reached a Shaker pharmacy. Even the man who introduced the yellow lady's slipper into formal American medical literature, Constantine Rafinesque, writing in 1828, admitted plainly that it was "long known to the Indians." The settler trade did not discover the plant's use. It took it. It scaled up a knowledge that already existed, and sold it. The flaunted yellow flower Blanchan admired was, in the same decades, being pulled out of the woods by the wagonload on the strength of what Indigenous peoples had known about it all along.

The hidden giant

Now we go south, and we go looking for the flower this whole chapter has been circling.

Somewhere in the bottomland woods of Arkansas, along a stream terrace where the spring seep keeps the ground wet at the foot of a slope, under beech and oak and tulip poplar, there is a lady's slipper standing nearly waist-high with a pouch the size of a goose egg. It is not yellow. It is ivory, cream, the palest butter going toward white, but never the bright gold of the northern yellows and never the snow-white of the small white slipper and never a breath of pink. The petals corkscrew off to the sides, longer than any other lady's slipper's on the continent, twisting like the yellow one's but stretched to extravagance. And the pouch — the pouch is the headline. It is the largest pouch of any lady's slipper anywhere on Earth, in any of the dozens of species across three continents. The biggest slipper in the genus is this pale southern one.

This is the Kentucky lady's slipper, *Cypripedium kentuckiense*, also, and better, called the ivory lady's slipper. It is tied with the showy lady's slipper for the tallest native slipper we have, and it holds the world record for the single largest flower in the genus outright. And for most of the time botanists have been walking American woods, they did not know it was there.

That is the part worth slowing down on. This is not some obscure little plant tucked into a crevice. It is the biggest, one of the showiest lady's slippers on the planet, growing in the accessible bottomlands of the American South, and it hid in plain sight for something like a hundred and fifty years. It hid by being too similar to something common. Every botanist who came across one saw a yellow lady's slipper, variety *pubescens*, the big yellow, that had simply grown unusually large and unusually pale, an oversized specimen of a plant they already knew. They had a box for it, and they put it in the box, and the box was wrong.

The unhiding took decades. In 1951 an amateur, an Ohio man named Jim Daulton, was on a fishing trip in Kentucky and noticed something off about the big slippers: too big, too pale, not quite right. He was, it turned out, the first to really see it. It took thirty more years for the formal recognition to catch up. In 1981 the botanist Clyde Reed published the description that finally named it a species of its own, *C. kentuckiense*, the most recently described of all our native lady's slippers. And then, because a claim like "this is a new species, not just a big variety" invites a fight, the genetics had to settle it. In 1998 a team led by Martha Case ran the plant's enzymes against the big yellow's and found it genuinely distinct: low genetic diversity and a unique signature, the fingerprint of a species that split off recently through a narrow founding bottleneck. Later microsatellite work confirmed it again. The ivory lady's slipper is real, it is its own species, and it is recently diverged, which is precisely why it was so easy to mistake. It split off the yellows so recently, in evolutionary terms, that it still looks like one of them, and the exact position on the family tree remains unsettled.

There is a small geography joke folded into the name, and it is worth the smile. It is called the *Kentucky* lady's slipper because Reed described it from a Kentucky plant. But Kentucky is not where it is most at home. The center of gravity of the whole species, the heart of its range, is well to the southwest, in the Ouachita Mountains of Arkansas, where it grows in its greatest numbers. The largest slipper orchid in the world is named for a state that is, more or less, its northern edge. The name honors the place it was noticed, not the place it lives.

And it lives apart from the northern yellows in part because it took a different kind of ground. Remember that the yellow lady's slipper is a lime-lover, a plant of sweet calcareous soil. The ivory one, *C. kentuckiense,* mostly is not. Across much of its range it grows in acidic, sandy, well-drained alluvium, the bottomland of southern hardwood forests, the seepy foot of a slope, ground that is wet from below but sour rather than chalky. It is the one native lady's slipper that made its home in acidic bottomland-hardwood country, which is a large part of why it lives down south on its own, in its own woods, instead of among the yellows it so closely resembles.

So how do you call it, with a big pale pouch in your hand? Three things settle it. First the color: an ivory that never warms to gold, a cream going toward white, where the big yellow is honestly yellow, true butter and gold. Second the size. The pouch is simply too big for any northern yellow, larger than any other slipper on Earth, past the upper limit of what a *pubescens* can grow. Third, and this is the one Sheviak used to draw the line, the opening itself is built differently: the mouth of the pouch replaces the bottom of its front face, and the sides of the slipper end there abruptly, without the inward curve the yellow one keeps. Ivory, oversized, blunt-mouthed. Now you can name it.

*The Kentucky lady's slipper (*Cypripedium kentuckiense*), the largest-flowered native, its ivory pouch the size of a goose egg, here in the Arkansas bottomlands at the heart of its range.*

The flower that may bend the rule

I told you, a couple of chapters back, the genus's central trick: every lady's slipper is a one-way trap that does its outcrossing for it. The rule held for every species we examined. The pink one runs the trap with a queen bumblebee at the front slit. The yellow one runs it with a sweat bee. The showy one runs it with a mid-sized leaf-cutter at the rear exit. Twelve flowers, one machine.

The ivory lady's slipper may be the one that bends it.

Watch what happens when a bee enters that enormous pouch, because biologists have. The bees that work it are digger bees, the genus

Anthophora, and the first oddity is that the males come too, and may help pollinate, which almost never happens elsewhere in the genus, where it is reliably the females that do the work. But the stranger thing is what the bees do once they are inside. In a tighter pouch a trapped bee is hustled along, scrabbling for the only exit, in and out in a hurry on the flower's terms. In this one, the biggest chamber in the family, with a roomier and looser exit canal, the bees linger. They do not panic and bolt. They have space. They have even been watched chewing on the lining of the pouch, which is not the behavior of an animal in a tight one-way trap fighting to get free.

So the ivory lady's slipper is the closest thing the genus has to an exception to its own iron rule. I will not overclaim it. The flower is still nectarless, still a deceiver, still loosely the same machine as its cousins, and nobody is saying it has abandoned the trap the way the self-pollinating sparrow's-egg slipper we met earlier abandoned the bee entirely. But the strict, ruthless, hustle-them-through kettle that defines the family is, in the biggest pouch of all, relaxed. The flower that broke the size record may also be the flower that bent the rule. The largest slipper in the world is, fittingly, the one least willing to follow the family business to the letter.

A secure flower and a vanishing one

I have been telling you these two are easy to confuse, and at the level of the flower they are. At the level of their survival they could hardly be more different, and the difference is the last thing I want you to carry out of this chapter.

The yellow lady's slipper is secure. On the standard global rankings it sits in the most reassuring category there is, a plant in no real danger of vanishing from the world: too widespread, too abundant across too much of a continent to be lost. It has its local troubles, as any plant does, and it is protected by statute in a good many states where it has grown scarce. But the species as a whole is in no danger. The most widely distributed lady's slipper we have is also one of the safest.

The ivory lady's slipper is not safe. It is ranked vulnerable, a global rank that means real risk of disappearing, and the numbers behind that rank are sobering. There are, by the best estimates, fewer than about three thousand of these plants left in the wild, anywhere, total. About half of the sites where it was known to grow have been lost in roughly the last twenty years. The largest slipper orchid in the world, the one botanists only properly recognized within living memory, is slipping away almost as fast as we learned to see it.

One of those losses is specific enough to lodge in the mind. In Knox County, Kentucky, a population of more than five hundred ivory lady's slippers, five hundred of a plant of which fewer than three thousand exist on the entire planet, was destroyed to build a Walmart. Stand those two facts next to each other. A flower so newly understood that the ink on its species description is younger than many of the people reading this. And a single parking lot that erased a sixth of all the plants that might exist. We finished learning what it was, very nearly in time to watch it go.

There is recovery work, hopeful and real, but it belongs to the chapter on mending what we have broken. For now I only want the two flowers held side by side: the common yellow that is safe, and the great pale southern twin that is not, the two of them so alike that for a century and a half nobody could tell them apart, and so unlike in their fortunes that one is everywhere and the other is nearly gone.

In the bottomland

I have never found a wild ivory lady's slipper. It is a plant of the southern bottomlands, the Ouachita seeps and the Gulf-woods stream terraces, a thousand miles from my own acid ridges, and I have not made that trip. The largest slipper orchid on Earth, and I know it the way you may have to know it, from photographs and other people's accounts. So let me build it for you out of those, and leave it standing where it lives.

Somewhere in a wet southern wood, at the foot of a slope where a spring seep keeps the ground soft, one stands nearly waist-high: a pouch the size of a goose egg, pale as cream and never gold, the long petals corkscrewing off to either side and browning a little at the tips.

From a few feet away, if you did not know, you would call it a yellow lady's slipper that had gone pale and grown large. That is what everyone called it, for a hundred and fifty years. Up close, knowing what to look for, you could finally name it: the ivory, the impossible size, the blunt mouth of the pouch. And you would be standing in front of a flower still, even now, in the middle of becoming a separate thing from the one it was mistaken for.

Down in that wet ground a fungus holds its roots and passes it carbon, the way it has done for longer than there have been botanists to be confused by it. The largest slipper orchid in the world goes on doing exactly what it does, swaying a little in a wind I will never feel, in a bottomland where almost no one will ever come to see it.

The Queen

The most beautiful wild orchid on the continent will, quietly, fight back against the hand that reaches for it.

Not with a thorn. There is no thorn anywhere on it, nothing sharp, nothing that draws blood. The showy lady's slipper is the tallest, whitest, most flagrantly regal flower in this whole book, a pink-and-white pouch held nearly waist-high in a sunlit fen, and the first thing most people want to do when they finally meet one is put a hand on it. A finger run up the stem is where the trouble starts, because the whole plant is furred. Stem, leaves, the big leafy bract behind the flower, every surface is covered in fine soft hairs, and a great many of those hairs are tipped with tiny glands that weep a faint sticky secretion. To the eye it reads as a kind of bloom, a velvet softness on a flower already lavish with it. To the skin of some people it reads as poison ivy. Pick an armful of this orchid and carry it home, and your hands may come away stained and your forearms broken out in a rash like a bad brush with *Rhus*. The most crowned, most painted, most gathered of our native orchids is also, in a small and serious way, defended. The queen does not like to be handled.

That is the surprise I want to carry through this chapter, because almost everything else about this flower runs the other way: toward show, toward fame, toward a kind of effortless supremacy nothing else in the genus can touch.

This is the showy lady's slipper (*Cypripedium reginae*). You have met it already in pieces across this book, a sentence here and a paragraph there. Now I want to give you the whole queen.

The showy lady's slipper (Cypripedium reginae)*, its broad white sepals and petals framing a pouch washed with rose. The queen of the native slippers.*

Why she earned the name

Start with the plain fact of her size. Stand a showy lady's slipper next to any of its eleven North American cousins and it overtops them. The stem rises somewhere between twenty-one and ninety centimeters (about eight to thirty-five inches), now and then breaking a full meter, which makes it the tallest native lady's slipper we have, sharing the honor only with the great ivory southerner of the last chapter. But there is more plant here than height. Run your eye up the stem and count the leaves.

This is the field mark. It is almost foolproof. Where the pink one we just left carries two leaves flat on the ground and a bare pole above them, and the yellows run three to five up the stalk, the showy lady's slipper runs leaves up its entire stem, three to nine of them, four to seven on a typical plant, the highest leaf count of any lady's slipper on the continent. They are big leaves, broadly oval, pleated with parallel ribs, clasping the stem and climbing it like rungs. A leafy *Cypripedium* with six or more leaves marching up a tall stem in an eastern fen is, essentially, this and nothing else. You can name it from across the wetland before you are near enough to see the flower at all.

Then there is the flower, which is the reason for everything. The pouch is large and round-mouthed, inflated like a small urn, and its ground color is white, washed and flushed with pink, sometimes a soft rose, sometimes a deep crimson-magenta laid on in veins and blushes that no two plants wear quite alike. The white is true white. The pink looks brushed on over it, soaking into the white the way color soaks into wet paper. Above and behind the pouch the sepals and the two lateral petals spread flat and pure white, broad and open, and here is a quiet thing worth noticing: those side petals do not corkscrew. The yellows twist their petals away to either side like curled ribbon. The showy one does not; its petals lie broad and flat, framing the pouch like an open hand. And tucked at the throat of the flower, guarding the way in, sits the staminode, the sterile shield that every slipper orchid carries in place of one of its stamens. On the showy lady's slipper it is its own small painting, white, marked with bright yellow streaks and a scatter of purple dots, distinctive enough that on a doubtful plant it alone will settle the question.

One flower would be enough. This plant often carries two, sometimes three, at the top of a single stem, which is rare in the genus and, paired with the height and the leaf count, all but unmistakable. Picture it. A flower taller than your knee, leafy to the top, crowned with two great white-and-rose pouches, standing in the wet sun of an open fen. Somebody had to name this plant, and when Thomas Walter did, in 1788, he reached for the obvious word. *Reginae*. Of the queen.

There is even a rare pure-white version, the pink drained entirely out of the pouch, all that watercolor wiped clean to leave a flower of unbroken ivory and snow. Botanists call it forma *albolabium*. To come on one standing among the ordinary pink-washed plants is to feel you have found the rarest thing in a place already full of rare things.

And she has a smell. Put your face near an open pouch on a warm June morning and a fragrance comes off it, faint and clean, a sweetness the people who have tried to pin it down keep likening to bluets, those tiny pale flowers of the spring woods. Here is the strange part. For all her fame, nobody has ever taken that scent into a laboratory and named the molecules in it. Not once. This is the celebrated state flower of Minnesota, the most painted orchid on the continent, and her perfume has never once been run through an instrument and written down. After the pink one, it is the largest such blank in the whole genus.

The name, and the name it replaced

A flower this conspicuous gets named more than once, and the showy lady's slipper has carried a small tangle of names worth one short detour, because it says something about how badly everyone wanted to do this flower justice.

Thomas Walter named it first. He was an English-born planter and botanist in colonial South Carolina, and in 1788 he published the first flora of North America written entirely in the Linnaean system of two-word Latin names, his *Flora Caroliniana*. On page 222 he set down *Cypripedium reginae*, the queen's lady's slipper, and died about six months later, having given the most magnificent orchid on the continent the one name it most deserved. By the rules of botanical naming, the first valid name wins, and Walter's was first.

And yet for roughly a hundred years, most botanists called it something else. In 1791 the English botanist Richard Salisbury named the same plant *Cypripedium spectabile,* the spectacular lady's slipper, and that name, *spectabile,* is the one that filled the nineteenth-century books, the European garden catalogs, the great color plates. Walter's "queen" sat in the record with priority and was largely ignored; Salisbury's "spectacular" did the work. It was not until the very end of the eighteen-hundreds that American botanists, sorting out who had named what first, quietly restored Walter's *reginae* to its rightful place, and *spectabile* slid into the synonymy where it sits today, a discarded second name. Two different botanists, a few years apart, each looked at this flower and named it for being too much. One called it the queen. The other called it spectacular.

The state flower named wrong, twice

If the naming history is a quiet comedy, what happened in Minnesota is a loud one, and it is my favorite thing about this flower's life among people.

Minnesota wanted a state flower. In 1893 the legislature obliged and named one: *Cypripedium calceolus,* a lady's slipper. There was only one problem, and it was a large one. *C. calceolus* is the European yellow lady's slipper. It does not grow in Minnesota. It does not grow anywhere in North America. The state had just made its official flower a plant that lived on the wrong continent.

How this happened is itself a small lesson in the confusion that runs through this whole genus. In the middle of the nineteenth century many botanists treated the American yellow lady's slippers not as their own species but as mere varieties of the European *calceolus*. So when somebody in 1893 reached for the Latin name of "the lady's slipper," the books handed them *calceolus,* and into the statute it went. For nearly a decade Minnesota's state flower was, on paper, a species no Minnesotan had ever seen growing wild.

Then a few people noticed. On the second of February, 1902, the *Minneapolis Tribune* ran a piece under a headline that says everything: "State Flower a Fake." And the prod that put the matter before the legislature did not come from botanists or schoolchildren, as the story is sometimes told. It came from a women's club, the St. Anthony Study Circle of Minneapolis, who took up the embarrassment and pushed for a fix. A state senator introduced the corrective resolution, and on the nineteenth of February, 1902, Minnesota struck the European impostor and named, in its place, a flower that actually grew there: the showy lady's slipper.

THE STATE FLOWE

This question of what rea species of the state flower i a grave one. The legislature, quest of the St. Anthony Stud Minneapolis, is trying to chang designated in the joint resolut 4, 1893, which reads:

Be it resolved by the Senate, t Representatives concurring, Tha lady-slipper," or moccasin flow dium calceolus, be and the sam designated and adopted as the st floral emblem of the state of Mi

The ladies of St. Anthony S assert that "calceolus" does r Minnesota and that the familia white lady's-slipper was the fl mind by the original petitio state flower, (the state flower of the Woman's Board of Wo Managers), the proper name "Cypripedium reginae."

Two of our contemporarie foul of this state flower que Tribune has the lady's-slipper the prairies, where it does nc the contrary, it has fondness f wooded places. The Tribune badly mixed in its Latin. undertakes to rebuke the Trib also entangled by the wiles familiar Latin word. The covers that the state manual "Cypripedium arietinum" is flower. This it indirectly d a cut of the flower it contents the designation "Cypripedium casin Flower, State Flower of Nothing is said about the s in the article devoted to the s which is equally non-committ precise designation of the sta says that the "flower shown companying plate" is " arietinum." But the law reads "Cypripedium calceolus law is above the blue boo cation. It is equally that "Cypripedium reginae" most people take to be flower—the pink and white lad Naturally the good ladies of thony Study Circle, which is for these editorial errors, in cipitated the subject, are stirred up. One of the mem

The 1902 Minneapolis Journal *editorial on the state-flower mix-up. Minnesota had accidentally adopted a European orchid; the error was caught and set right. Public domain.*

The correction has one more delightful wrinkle. The statute did not name the plant *Cypripedium reginae* and leave it at that. It named, in plain English, "the pink and white lady slipper," and that homely four-word phrase, no apostrophe, no hyphen, is to this day the legal name of Minnesota's state flower. A flower the books had argued over in two competing Latin superlatives became, in law, simply the pink and white one. And the state has guarded it since: a 1925 act made it illegal to dig or sell the showy lady's slipper, or indeed any native orchid, anywhere in Minnesota, one of the earliest plant-protection laws of its kind in the country. The queen, finally crowned, under an everyday word.

She wore one other crown, briefly, and lost it. Prince Edward Island made the showy lady's slipper its provincial flower in 1947, then took the title away in 1965 and handed it to the common pink one of the roadside ditches, because the showy queen had grown too rare on the island to serve as anyone's emblem. We have already stood in those ditches from the pink one's side; I only note the loss here from the queen's. The most magnificent flower the island had was demoted for being too hard to find. Rarity is a poor qualification for a flag.

The queen of the cypripediums

The fame is not only official. This flower has lived in the language of people who loved it for as long as Americans have written about wildflowers, and one passage gave the showy lady's slipper the title that has followed it ever since.

In the eighteen-nineties, John Burroughs was the most widely read nature writer in America, a bearded sage of the Hudson Valley whose essays sent a whole generation into the woods looking. He knew the showy lady's slipper, and he knew where a colony of it stood in a swamp near his home, and he guarded the spot the way people guard such things. In his 1894 book *Riverby* he tells of a friend, an orchid enthusiast, who "came one June day a long way by rail to see this flower." Burroughs walked him to the edge of the swamp. Then, in a gesture I have never been able to forget, he "lifted up the branches as I would a curtain, and said, 'There they are.'" And of the flower behind the curtain he wrote the words that stuck: that nearly every June his friend still came "to feast his eyes upon this queen of the cypripediums."

This queen of the cypripediums. It is exactly right, and it is the phrase the flower has worn for more than a century since. Burroughs understood something the state legislatures and the rule-bound Latin only gestured at: that you do not stumble on this flower the way you stumble on the pink one in a ditch. You make a pilgrimage to it. You come a long way by rail. Somebody who knows the secret swamp lifts a branch like a curtain, and there she is. The pink one is the orchid you find. The showy one is the orchid you are shown.

Eloise Butler, who built one of the first wildflower gardens in the country in Minneapolis, said the same thing more plainly a few years later. Writing in the *Minneapolis Tribune* in 1911, with the state-flower question freshly settled in the showy one's favor, she called it "the greatest prize of the swamp," her state flower, "the pink and white Lady's-slipper." Across thirty years, two careful observers reached for the same register: not affection, the way one feels affection for the pink one, but something closer to homage.

The defense hidden in plain sight

And here is where the chapter turns back to where it began, because all that homage was paid to a flower that quietly resists the very people paying it.

Look closer than the homage usually gets, at the soft fur the opener warned you about. Under a lens, a great many of those hairs turn out to be **glandular**, each one tipped with a tiny gland that weeps a faint sticky fluid. The velvet softness is part of why the flower reads as so lush. It is also the problem. Those glandular hairs carry an irritant, and on the skin of some people it raises a contact rash much like poison ivy's, redness and itching and small blisters where the plant touched. Not everyone reacts. But enough people do, and badly enough, that the showy lady's slipper holds an odd distinction in this genus: it is the orchid the doctors wrote up.

The story is documented, and the first telling is wonderfully specific. In 1875 a Chicago botanist named Henry Holmes Babcock, a man who helped organize the city's early botanic garden, gathered an armful of showy lady's slippers, then known by the name *spectabile,* and paid for it. The glandular secretion stained his hands. Within a day or so his face had broken out so badly that, in his own account, "his whole face presented the appearance of a severe case of *Rhus* poisoning."

Rhus is the old name for poison ivy and its relatives. He wrote it up for a Chicago pharmacy journal, and it stands as the earliest American report of contact dermatitis from a lady's slipper. About twenty years later another botanist, Daniel Trembly MacDougal, did what curious botanists do and tested it on himself on purpose, to find out exactly how the plant raised a rash.

What is in the hairs has a name, though here I have to be careful, because the careful version is the true one. The irritant compound has been isolated and identified from the European yellow lady's slipper, where chemists called it cypripedin, and on the strength of the family resemblance it is reasonably presumed to be in the showy lady's slipper's glandular hairs too. But here is the honest gap. No one has actually pulled the compound out of *C. reginae* itself and confirmed it in the laboratory. The dermatitis is real and old and well attested; the exact chemistry inside this particular plant is inferred from its relatives, not yet demonstrated. I would rather tell you that plainly than claim more than the record holds.

Now hold the whole shape of it. The showiest flower in the genus, the one we crowned twice in Latin and made a state flower and made a pilgrimage to, is sheathed head to foot in a soft glandular coat that bites a certain fraction of the hands that reach for it. I find that quietly marvelous. We have a long habit, in this book, of discovering that a lady's slipper's beauty is doing a job — luring a bee, working a trap. Here is a different job, worn on the outside, in plain sight, mistaken for ornament: a defense disguised as softness. She is gorgeous and she is armed.

I want to be clear about what this is and is not. It is dermatology and a thread of nineteenth-century history, not an endorsement. The showy lady's slipper was only a minor bit player in the old "American Valerian" root trade, far behind the yellows and the pink one, and the rash-raising foliage was part of why the root-diggers tended to leave it alone. This is not a medicine to make anything of. It is a flower with a coat that defends it, and a doctor's note from 1875 to prove it.

What you already know about her

Everything to this point has been the queen's public life: her names, her titles, her small armed secret. Her private biology you have largely met already, in earlier chapters, and I will not re-reveal it; I only want to gather the threads so the whole plant stands in one place.

You already know she is slow. This is the flower I used, a few chapters back, to show you the patience buried in this genus: from a dust seed to the first open pouch takes the showy lady's slipper something like fourteen to sixteen years, the longest stretch documented for any lady's slipper on the continent, almost all of it underground and unwatched. You know, too, that she can vanish: a plant which appeared one June may decline to come up the next, riding out a year or several below ground before reappearing, so that the careful people who count her have learned never to call a missing plant dead. The stem you find may be twenty to fifty years old. The buried individual beneath it may be far older.

You know who works her flower, because we watched it happen. The round mouth of her pouch is generous, an open door any bee can enter, but the two small rear exits behind the column are narrow, and they sort her visitors by size on the way out: admitting the mid-sized leaf-cutter bees that carry her pollen, turning away the bumblebee queen who blunders in and can only back out the way she came. The pink one sieves her bees at the front door, by force. The showy one sieves hers at the back, by fit. The trap and its twin.

And you know whose company she keeps underground. In the black-ash and cedar fens of the far Northeast, the showy lady's slipper shares dozens of fungal partners with the black ash trees standing over her, plugged into the same buried network. Which is why, of all our orchids, she is the one most quietly threatened by an insect that has never touched her: the emerald ash borer killing the ash, and through the ash, perhaps, thinning the fungal web she lives by. I told that story where it belonged. I raise it here only so you remember that the queen does not stand alone. The most independent-looking flower in the genus is as plugged into the buried web as any of them.

There is one habitat fact worth a clean clause, because it mirrors the chapter before this one. The pink one took the sour ground. The showy one took the sweet. She is the most lime-loving lady's slipper we have, a creature of calcareous fens and cedar swamps where the water moves through old seabeds and comes up rich in calcium, ground a pink lady's slipper could not survive in for a season. The two queens of this book stand at opposite ends of the same chemistry. But I will leave the larger story of how the whole genus sorts itself along that line for later; it is its own chapter.

Common nowhere near everywhere

So she is magnificent, and slow, and particular, and plugged into a fen. What does that add up to for her future? A complicated answer, and a useful one, because the showy lady's slipper is the textbook case of a conservation problem the simple labels hide.

On the global rankings she sits in a reassuring tier, "apparently secure," and at the scale of the whole continent that is honest. She ranges across some twenty-five states and most of eastern Canada, and her strongholds in Ontario and Minnesota and the upper Great Lakes hold her in real numbers. There is no near-term prospect of losing the species. But that rank is a verdict about the whole range at once. A range is not a place. Walk to the edges of hers and the picture darkens fast. She is listed as endangered, or its legal equivalent, in something like nine states: Arkansas, Connecticut, Illinois, New Hampshire, New Jersey, Pennsylvania, Tennessee, Virginia, West Virginia. She is threatened or vulnerable in several more. And in Kentucky and Maryland she is reckoned simply gone. A flower can be secure across a continent and vanishing off both its edges at once, and this one is. She is common nowhere near everywhere she once reigned.

The pressures on her are the ordinary, sobering list. Her fens get drained and built on; she is exquisitely sensitive to anything that changes how the water moves, and a wetland filled or a water table dropped takes her with it. Deer browse her hard. And because she is the most iconic native orchid we have, she is dug, a documented target of the online trade in wild plants, lifted out of the ground by people who want the queen in their own garden and almost never get her to live there, since a transplanted showy lady's slipper, severed from her fen and her fungus, mostly dies.

She has one stranger threat, and it is almost a dark joke on her whole biology. A small European butterfly, the European skipper, was introduced to Ontario around 1910 and has spread widely since. It is drawn to the showy lady's slipper's pouch, and it gets *stuck* in it: trapped in the same chamber built to sort bees, but with no pollen to carry, a small wrong-shaped body clogging the works. Whether enough skippers get caught to actually block the leaf-cutter bees she depends on is not yet known. But picture it. The trap that made her, the elegant size-sieve we admired, jammed by an immigrant insect that does her no good and may do her harm. The mechanism that sorts her true partners cannot tell a useless visitor to stay out.

I will not pull the full reckoning forward into this chapter; what we owe these flowers, and how the mending is going, belongs to the end of the book. For now I only want the paradox left standing. The most celebrated lady's slipper on the continent is also the one losing ground at the edges of her kingdom, fen by drained fen.

The asymmetry

There is one more thing I owe this flower before I leave her, and it is an absence.

This orchid grows across the heart of Anishinaabe homelands. It is the state flower of Minnesota, a state whose very name is an Anishinaabe-derived word. And yet the showy lady's slipper has, in the published record I can reach, the thinnest documented Indigenous-language naming of any of the eastern lady's slippers in this book. For the pink one I could set down the moccasin flower, *makizin* carried in Anishinaabemowin, with named sources and a living tradition behind it. For the yellow one I could set down a Cherokee name recorded in the eighteen-nineties. For this one, the spectacular state flower standing in Anishinaabe country, the trail in the published sources goes faint.

There is a tentative name. The ethnographer Frances Densmore, working among the Ojibwe in the nineteen-twenties, set down an Anishinaabemowin word for the showy lady's slipper, *Ago'bisowin,* which she glossed simply as "lady's slipper." A later compilation reads the word more fully, as something like "a remedy tied on a wound," though that fuller gloss reaches me secondhand, not from Densmore herself. I set the name down carefully and with a flag, because what holds it up is thin: it rests on a single outside ethnographer's record from a century ago, and the most authoritative regional institutions have not, as far as I can find, confirmed it for this species. So I name it, I name where it seems to come from, and I build nothing on it. *Ago'bisowin,* perhaps; a remedy tied on a wound, perhaps. Held loosely, pending people who would know far better than I do.

I want to name that gap as a finding rather than paper over it. There may well be a deeper relationship here that the published record simply never captured. The blankness in my sources is not the same thing as a blankness in the world; it is a measure of what the settler record failed to write down, or never had the right to.

The showy lady's slipper's staminode: a sterile shield, white with yellow streaks and purple dots, that helps steer a trapped bee toward the exits.

In the fen

I came on a colony of them once, in a calcareous fen in the upper Midwest, in late June, in the years the Coast Guard had me on the Great Lakes. No one lifted a curtain for me; I had walked in over the quaking sedge mat with wet feet and found them on my own. There were a few dozen, scattered through the open part of the fen where the sun came down hard, each stem leafy to the top, most of them carrying two pouches, white blushed with rose, standing taller than I always expect them to. The light was full on them. They looked exactly as regal as everything written about them promises.

I crouched at the edge of the colony. I did not touch them, partly out of habit and partly because I knew better. A leaf-cutter bee was working one of the nearer pouches, in through the round front mouth and, after a while, out one of the narrow rear doors, dusted and used and free. Somewhere under the sedge mat, in calcium-sweet water I could not see, the fungus she lives by held the roots of plants that might have been standing here since before I was born, doing the quiet work it has always done. I thought about the armful of these a Chicago botanist gathered a hundred and fifty years ago, and the rash that taught him what the softness was for, and I kept my hands to myself. The tallest flowers nodded a little in a wind I could feel now, off the open water, and the pink-washed pouches swung on their leafy stems, and I sat in the wet ground and looked at the queen for a long time. When I finally rose, the colony was exactly as I had found it.

The Prairie and the Ram

The same week in early June, in two places that could hardly be less alike, the two smallest lady's slippers on the continent are in bloom.

One is out in the open, in the sun, in a remnant scrap of wet tallgrass prairie somewhere in the upper Midwest, where the big bluestem is still short this early and the ground underfoot is alkaline and damp. Down among the grass, on a plant no taller than a ruler, hangs a tiny pouch of pure ivory white, the only truly white slipper we have. The other is far to the north and in deep shade, in a cold northern white-cedar swamp where the moss stays wet and the air under the conifers is chill even at midday. There, on a plant just as small, hangs a stranger thing: a little white pouch netted over with red veins, pointed sharply downward, with two narrow sepals sticking out to either side like horns. Two flowers, two opposite worlds, both barely knee-high to the plants around them.

Here is the odd fact I want to start with. For two flowers this distinctive, this strange, this rare, almost nobody wrote anything down. No moccasin tale attaches to either one the way it does to the pink lady's slipper. Neither is anybody's state flower. When the nineteenth-century medicine trade was pulling lady's slippers out of the eastern woods by the wagonload to sell as a nerve tonic, it passed both of these by. There is scarcely a poem. The eastern pink one lives in folklore and the showy one lives in legislation, and these two, the oddest and rarest natives of all, left so faint a trace that you could read a shelf of old wildflower books and never know they existed.

That silence turns out to be worth listening to. But first the flowers, both small marvels built around the same surprising idea.

Two flowers nobody planned together

These two are not close relatives. Not the same branch of the family, not the same habitat, not the same pollinator, and as far as anyone knows they have never grown side by side in the wild. By every ordinary measure they belong in separate chapters. I have put them together for the one thing nothing else in the genus shares: they are the two smallest lady's slippers in North America, small in a way that turns out to be the key to both their lives.

Start with the small white lady's slipper (*Cypripedium candidum*). The plant stands twenty to thirty-six centimeters tall (eight to fourteen inches), with three to five pleated leaves up a slender stem and, at the top, usually a single flower. The pouch is the headline. It is pure ivory white, faintly purple-veined inside near the base, and tiny, fifteen to twenty-seven millimeters long, the smallest white pouch of any lady's slipper anywhere. And it is genuinely white, which no other native slipper is. Others carry white in the pouch, but always flushed with pink or purple; this one alone is clean white, the color of a candle. The Latin says exactly that. *Candidum* means shining white, and the botanist who attached it in 1805 was naming the one unmistakable thing about the flower.

Now the ram's-head lady's slipper (*Cypripedium arietinum*), small in the same league, ten to thirty-five centimeters (four to fourteen inches), with three or four blue-green leaves up the stem and, again, a single flower. But where the white one is plain and pure, the ram's-head is the strangest-looking lady's slipper on the continent, and it breaks the family's rules in two ways at once that no other native does.

Look first at the sepals. On essentially every lady's slipper the two lower sepals are fused into a single structure behind and below the pouch, so seamlessly joined that we gave the fused thing its own name, the synsepal. The ram's-head does not do it. Its two lower sepals are free, separate to the base, spreading to either side like a pair of narrow horns. It is the only native lady's slipper that leaves them unfused, the one flower in the genus with no synsepal at all.

Then the pouch. It is not the inflated slipper you have come to expect: small, ten to sixteen millimeters, conical rather than rounded, white, netted with fine reddish veins, furred with soft white hairs. And instead of bulging forward it points downward, the tip deflected sharply toward the ground, so it hangs like a little muzzle. Put the down-pointed conical pouch together with the two horn-like sepals and a face appears: the blunt, down-nosed, horned head of a ram. Whoever first looked hard at this flower saw it too. *Arietinum* means ram-like, from *aries, arietis,* a ram. It is the same word that names the chickpea, *Cicer arietinum,* whose little hooked seed was thought to resemble a ram's head. The flower and the bean carry the same animal in their names for the same reason.

A ram's-head lady's slipper (Cypripedium arietinum)*, among the smallest natives. Its pouch tapers to a downward point, and its two lower sepals stay free and spreading instead of fusing.*

And yet, for all they look nothing alike and live nowhere near each other, at the pollination they have arrived at the same solution. That solution is the heart of this chapter.

The narrow door

A few chapters back I showed you that a lady's slipper's pouch is not really a trap. It is a sieve: the bee enters one opening and is forced out a single narrow exit, and that exit is sized to admit a body of one rough dimension and turn every other away. Each species sets the size of its door, and in setting it, chooses its partner. These two flowers take it to the limit.

With the smallest pouches, they set the narrowest doors. So the bees that fit those doors are the smallest in the whole business, a guild of the tiniest bees that visit any lady's slipper on the continent. Not bumblebee queens, not the leaf-cutters that work the showy lady's slipper; both far too big to pass a pouch this small. The visitors are little mining bees and little sweat bees and small carpenter bees, insects so small most people would see a gnat or a fleck and look past them, and the flower has handed them its whole reproductive future.

The ram's-head does the same, and harder. Its pouch is smaller still, and the entrance is a slot only about one to two millimeters wide, narrow enough to filter the world down to one genus of very small bee. What passes through is tiny *Lasioglossum* sweat bees, one of them a metallic-blue species. The botanist Warren Stoutamire worked this out in 1967, and the picture has held. The ram's-head is one of only about four lady's slippers in the entire world pollinated by *Lasioglossum* and nothing else; its Asian sister is another of the four. And there is one final detail I love: once a ram's-head flower has been fertilized, its upper sepal droops down over the entrance. The flower that has done its work shuts up shop.

Here is the turn to carry out of this chapter. We keep calling these flowers small. The better word is *narrow*. Each has built its whole life around a narrow door, and not only at the moment of pollination. The white one lives behind a narrow door of habitat, one kind of ground alone. The ram's-head lives behind its own, a cold niche served by a single genus of bee through a slot barely a matchstick wide. Each flower is fitted to its world with a precision that leaves no room to spare.

But a narrow door is the most selective thing there is, and the most easily shut. Drain the fen, plow the prairie, warm the swamp, lose the one small bee, and there is no second door to slip through. These two are not weak. They are precise, and precision in a changing world is a kind of risk. That is why both are in trouble: each arrived, by itself, at the same beautiful, dangerous answer, and the answer is be the narrowest door.

The prairie flower the plow outran

The small white lady's slipper is the only lady's slipper in North America that is a true creature of grassland rather than forest — one of only a handful of native orchids anywhere on the continent that live in open prairie at all. It does not merely tolerate the prairie. It is an obligate of it, a plant of the wet tallgrass, the prairie fen, the open lime-rich ground that ran in an unbroken ocean across the middle of the continent before the steel plow came.

You can read the ground by it. Botanists who score prairie remnants for quality keep a list of plants that appear only in the genuine never-plowed article, and this flower gets the maximum. Find it growing and you are standing on original prairie that has never been turned. It does not colonize a restored field or a roadside or a recovering pasture; it marks the surviving ancient thing, almost gone now, which is exactly why it is so rare. It is an early bloomer too, late May into early June, a week to ten days ahead of the northern yellow fen plant it sometimes grows beside.

It lives on a thin margin even in a good year. The botanist John Curtis, counting in the nineteen-forties, found this flower setting about two-tenths of a fruit per flower, and even those few are not safe: on a prairie in northern Ohio, biologists watched the white *Misumena vatia* crab spider, camouflaged inside the pale pouch, ambush the very bees the flower needs at its own door.

A white crab spider, all but invisible against a pale pouch, waits in ambush for the very bees the flower depends on. (Shown here on a sparrow's-egg lady's slipper.)

But none of that is what nearly erased it. The plow did. When the tallgrass prairie was broken for farmland through the eighteen-hundreds, it was broken almost completely: by most reckonings about ninety-five percent of the original tallgrass is gone, one of the most thorough conversions of a major ecosystem anywhere on Earth. The small white lady's slipper, growing only there and never returning once the ground was turned, went down with it almost acre for acre.

And here the flower connects to the silence I opened with. Through those same decades the eastern yellow and pink lady's slippers were being dug out of the woods by the ton and sold as American Valerian, a nerve tonic listed for years in the official drug books, a real and damaging trade. The small white lady's slipper has essentially no place in it, for a grim and simple reason: by the time the root trade reached its peak, the prairie was already going under the plow, the white flower already disappearing with it. The diggers never got the chance. It was gone with the prairie before the trade could find it, lost not to the apothecary but to the farm.

Death by marriage

There is one more way this flower is being lost, the strangest of all, because the agent is not a plow or a spider or a trade but the yellow lady's slipper growing next to it. Where the small white lady's slipper still survives, it often grows within reach of the yellow we met a couple of chapters back, and the two will cross. Their hybrid has a name, *Cypripedium ×andrewsii*, an in-between flower with a pouch part white and part yellow and a smudge of magenta the white parent never had, and it is fertile enough to keep crossing back into one parent or the other, generation after generation. That fertility is the danger. When a small population of the white flower sits beside a large, vigorous population of the yellow, the crossing runs mostly one direction, abundant yellow genes flooding into the scarce white ones until the white flower's distinct identity dissolves into the hybrid swarm. In Norfolk County, Ontario, a population is documented to have been swamped exactly this way, genetically absorbed until, by about 1984, the pure white plant was gone from the site. Not plowed, not dug, not browsed. Bred out.

There is a piece of botanical history folded into that hybrid's name. One of its forms was first described in 1932 by a botanist who put his name to it as a nineteen-year-old undergraduate. His name was John Curtis, and you have met him: the patient Wisconsin botanist whose careful counting gave us the first honest measure of how slowly these orchids grow, the man who opened our chapter on longevity. Here he is at the other end of his life, barely out of boyhood, already bent over the small white lady's slipper and getting his start on the plants he would spend a career learning to read.

The flower that runs on the wrong fungus

No lady's slipper can live without its fungus. The dust seed cannot germinate alone; a fungal partner must thread into it and feed it through the long underground years before a first leaf ever shows. And the orchids are, as a rule, choosy: most partner one particular group, the Tulasnellaceae, the genus's standard companion. But the prairie has its own fungal economy. As a group, the rare prairie orchids mostly partner a different family, the *Ceratobasidium* group; in one combined set of more than a hundred fungal isolates pulled from prairie orchid roots, around eighty-five percent were this prairie partner, not the standard *Tulasnella*.

The small white lady's slipper is the exception. The handful of times anyone has sampled the fungus inside its roots, what they found was not the prairie's *Ceratobasidium* but *Tulasnella*, the genus's old ancestral partner, the one almost all the other prairie orchids gave up for the local family. It lives in the prairie's soil, surrounded by orchids running on the prairie fungus, and quietly keeps faith with the old partner instead. It is the fungal-network outsider, a *Tulasnella* orchid in a *Ceratobasidium* world.

The evidence under that is thin, and the thinness is part of the wonder. Fewer than ten small white lady's slipper plants have ever been sampled for their fungi anywhere on the planet; the whole striking picture rests on a handful of roots. And the cleanest open question sits right beside it: nobody has ever checked which fungus the white-and-yellow hybrid runs on. The white parent keeps the old partner, the yellow parent is a generalist, and what their offspring inherits underground, in the one place it most matters, has never once been looked at. It waits in the prairie soil for someone with a microscope and some patience.

The flower the law caught

The small white lady's slipper holds a distinction no other lady's slipper on the continent does, a melancholy one. It is the only lady's slipper in North America that a national government has formally listed as a species at risk. In Canada it is listed under the federal Species at Risk Act, assessed as Threatened. The white flower is the one the law finally reached for, and the reason is everything we have just walked through.

The largest single population left in the world is on protected prairie in Manitoba, and the counts there tell the story. In one census more than thirty-four thousand stems were tallied. The population climbed for a while, to about fifty-five thousand by 2010, and then, watched carefully over twenty years, began a documented decline, pressed by deer browsing, by woody shrubs creeping in over the open ground, by drought. Even the stronghold, tended and counted and protected, is sliding downward.

I want to set one more thing beside the law, gently and with a flag on it, because it is a different kind of custody than a statute. In Ontario, more than ninety-five percent of all the small white lady's slipper that survives grows on the land of a single First Nation, the Walpole Island First Nation, on Bkejwanong Territory, an Anishinaabe community whose Heritage Centre stewards some of the finest unplowed tallgrass prairie remaining in the country. Almost the entire Ontario population of this flower persists because that land was never broken, and it was never broken because it has been held and cared for by the people whose land it is. This is not conservation in the sense of a designation drawn on a map by an agency. It is custody, older and more direct. A flower surviving on a people's own ground, kept by their keeping of it. I name it, point to it, and go no further, because that relationship is theirs to describe, not mine.

The ram in the cold swamp

Now leave the prairie and the sun and go north into the shade, because the other flower's strangeness deserves a full look.

The ram's-head is a flower of cold places, and that is the most important thing about where it grows. People call it a lime-lover, and it is true you usually find it on base-rich, sweet ground. But the substrate is not the rule. The rule is cold. Within that one constant it turns up on a surprising variety of footing: dense northern white-cedar swamps and tamarack bogs and open fens on saturated peat; and, just as readily, dry conifer uplands of jack-pine and cedar over limestone-influenced sand and the cobble of old Great Lakes beaches, on the flat limestone pavements called alvars. The plants are biggest in the wet cedar swamps and scarcest there; smallest on the dry shore sand and there, oddly, most abundant, sometimes in colonies of thousands. Wet swamp or dry shore, the constant is the chill. The ram's-head is the cold-loving near-opposite of the pink lady's slipper that took the warm acid woods: where the pink one wants sour ground, this one wants cool air.

It blooms early, mid-May into early June, each flower lasting only about a week, and it is sweetly fragrant, a clean vanilla to those who have put their faces to it. Whether that is the same sweetness as the fen-yellow's vanilla or something all its own, nobody knows, because the ram's-head's scent, like the showy lady's slipper's, has never once been taken into a laboratory and analyzed.

And it is the loneliest twig on the family tree of any lady's slipper we have. Its oddness, those free sepals and the stamen-like shield in its throat, so unsettled taxonomists that they twice lifted it out of *Cypripedium* into a genus of its own, as *Criosanthes* in 1818 and as *Arietinum* in 1833. Each time the modern genetics folded it back home, because it does belong, however strange a member it is. But it sits alone there, the sole North American member of its own section, section *Arietinum,* the species the section is named for. Its single closest living relative is not on this continent at all: a small slipper of the Himalaya and southwestern China, *Cypripedium plectrochilum,* the only other member of their section in the world. The whole branch holds two species. They split from a common ancestor something like sixteen million years ago, in the middle Miocene, and have carried that ancient oddness on opposite sides of the planet ever since.

So the ram's-head is a kind of relic. It is one of two such oddities among our native lady's slippers, the one that broke the genus's floral template. Its odd twin, a western mountain species that broke the genus's leaf template instead, has Asian relatives of its own, and that one is a story for the next chapter. One last contrast before we leave the biology: the ram's-head must have its bee. It is self-compatible but cannot fertilize itself, so without that one small *Lasioglossum* through its narrow door, it sets no seed. The self-pollinating sparrow's-egg slipper needs no vector at all. Same cold country, opposite bets.

The mystery in its roots, and the map that points the wrong way

With the white flower, nobody had checked the hybrid's fungus. With the ram's-head the gap is bigger: nobody has resolved which fungus the *plant itself* runs on. Of all twelve native lady's slippers, it is the one whose underground partner is most completely unknown. There has been exactly one dedicated attempt, at the great Michigan dune population, and what grew from the cultured roots was not the expected orchid partner but a set of dark filamentous fungi that turn up in many plants, almost certainly an artifact of a dish that favored the wrong tenants. And that attempt sampled only the dry, sandy dune end of the range; the cold cedar-swamp end, where the plants are largest and the wet peat is a different world, has never been sampled at all. The honest answer is a blank. Of everything I cannot tell you in this book, this is the one I would most want to know.

The conservation picture is unusual too. Across all fourteen states and provinces in its range, the ram's-head is ranked imperiled to some degree, and in none of them is it considered secure. Not one. It is not listed under any national endangered-species law, unlike the white flower, but its rarity is range-wide and real. In Nova Scotia, where a far-flung outpost clings to gypsum barrens more than three hundred and thirty kilometers (about two hundred and five miles) from the nearest plant in Maine, the province lists it as endangered outright.

The flagship number is that Michigan dune population, holding more than ninety percent of all the ram's-head known to exist. It has been counted twice, fourteen years apart. Around 2009 the count came to roughly three and a half million stems. By 2023 it had fallen to about one and a quarter million, as the oldest jack-pine stands on the dunes closed their canopies and shaded out the orchid that needs open ground. The single greatest population of this flower on Earth lost two of every three stems in a decade and a half.

There is one more fact about its future, and it has to be set honestly against that decline. When a team modeled the climate-suitable range of six of our lady's slippers, five shifted north or northeast, the expected poleward direction; the ram's-head was the exception, its suitable habitat sliding west by about three hundred and thirty-five kilometers. Do not take comfort from that. The actual flower is rare everywhere it grows, gone or going from its southern edge, crashing in its single great stronghold. A band of suitable climate sliding sideways on a map is not a plant gaining ground, and a flower this narrowly fitted to cold, base-rich, open ground has nowhere easy to go.

The medicine trade has one last encounter with this flower, and it is pure cipher. The ram's-head was not dug and sold the way the eastern yellows were, but it was not entirely missed either. In one of the great old dispensatories of American medicine, the *King's American Dispensatory* of 1898, it is listed among the lady's slippers thought to share the medicinal properties of the others. Named, written down. Then never priced, never gathered. It is the plant the trade cataloged but never dug, fitting for a flower whose whole place in the human record is to be noted and passed over. Even the orchid hunters who covet it as a trophy mostly fail to keep it alive: it is among the very hardest of all native lady's slippers to grow in a garden, though it lives ten to sixteen years and longer in the wild.

The ones the story forgot

I said at the start that these two flowers left almost no trace in the human story. It is the truest thing the two of them share, truer even than their narrow doors, and it has to be handled with care.

For most of the lady's slippers in this book I have been able to set down names and stories from the peoples who knew these plants long before any botanist reached for Latin. The pink one carries the moccasin flower, in Anishinaabemowin, with named sources and a living tradition behind it. The yellow one carries a Cherokee name recorded in the eighteen-nineties. The showy one, even where the record thinned, carried at least a tentative word. The careful thing, the only honest thing, is to name those and defer to the people they belong to.

For these two flowers, that part of the record goes nearly silent. In the published sources I can responsibly reach, there is no species-specific Indigenous-language name set down for the small white lady's slipper, and none for the ram's-head; the small white one's record is the thinnest of any flower in this entire book. This does not mean the flowers went unknown or unvalued by the peoples in whose homelands they grew; it means the settler record did not set it down. And I can be fairly confident the silence is in the record and not in the world, because the very same sources that go quiet on these two are explicit and detailed about the others. The compilations that carefully record the moccasin flower for the pink one fall silent here. The gap is not random. It sits exactly where two flowers were rarest, oddest, and least likely to be commonly gathered, a prairie flower already vanishing under the plow and a boreal flower hiding in cold swamps. The blank in the settler sources is a real and meaningful finding, a measure of what was never written down, or never asked, or never anyone's to record in print.

So I will name the gap rather than fill it. I am not going to invent for either flower a folklore as rich as the moccasin flower's, to imply some parallel story because the better-known flowers have one. There may well be knowledge held about these plants that never reached print, and if there is, it is not mine to reach for or to guess at. The honest move is to say plainly that the story is quiet here, that the quiet is itself worth noticing, and that the right response is not to manufacture sound.

There is one nineteenth-century voice I can add, and only one, and it is small. The ram's-head's strangeness was at least seen and written about once, early, by a Canadian naturalist. Catharine Parr Traill, an English-born pioneer writer who produced the first popular Canadian wildflower book in 1868, looked at this flower and saw the animal in it that everyone since has seen. She described, in the ram's-head, the resemblance to "a sheep or ram, with the horns and ears." It is a settler naturalist's field note from a century and a half ago, written in the period's own language, and I quote only that one clause of it. But it is the earliest written record I can find of someone looking hard at this odd little flower and naming what they saw, and it is nearly the whole of the flower's life in the literature. One clause, about the horns and ears.

A small white lady's slipper (Cypripedium candidum)*, the only pure-white native, on a Kentucky hill prairie. Where it still grows, the ground was very likely never plowed.*

In the swamp

I have never stood in a ram's-head swamp. It keeps to cold northern ground I have not walked — the white-cedar swamps and tamarack bogs of the Great Lakes country, the alvar and dune shores far from my own acid hills. Of the two flowers in this chapter I have met neither in the wild: the small white one was plowed out of reach a century before

I was born, and the ram's-head I know only from photographs. They are, for me, what this whole chapter says they are for nearly everyone: more written about than met, and barely even written about.

I have it only as a picture, then. In a cedar swamp somewhere in the north, on wet moss over black peat, in air cold enough to show your breath on a summer morning, a ram's-head lady's slipper hangs on a stem no taller than your hand: the netted white pouch pointed down at the moss like a muzzle, the two narrow sepals held out to either side, the whole improbable ram's face in the green dark, sweet-smelling to anyone who bends close. It is small, and so shaded that an eye slides right over it. You would have to earn it, or go home without it.

And around it, silence. The silence I keep coming back to. No story I can tell you, no name from the people who walked this country first that ever reached the books I can read, scarcely a poem, one clause from one old naturalist about its horns and its ears. One of the rarest and strangest flowers on the continent, and almost nobody ever wrote it into the human record at all. It goes on anyway, doing exactly what it has done for millions of years, fitting its narrow door to its small cold world, swaying just slightly on its stem, where almost no one will ever come to see it.

The Western Three

Start with a number, because the number is the whole point. The mountain lady's slipper grows across something like three and a half million square kilometers of western North America. It turns up in the Alaska panhandle, in cool conifer shade near sixty-one degrees north on a limestone slope in the Wrangell–St. Elias backcountry. It turns up far south of there in the eastside Cascades, beside cool brooks in Yosemite, and east of all of it, improbably, as one isolated clump in the Cypress Hills of southwestern Saskatchewan, a lone outpost of lodge-pole pine on the dry plains with no other population for hundreds of miles. Ten states and provinces. An empire of a flower.

Now drive to one corner of the Pacific West, where southwest Oregon meets the northern California line, and go looking for the California lady's slipper. You will not find it across ten jurisdictions. You will find it in two, and not much of those two: a scatter of wet seeps in the Klamath–Siskiyou mountains, roughly thirteen counties in California and five in Oregon, and within those counties only the specific places where a particular greenish bedrock breaks the surface and groundwater trickles out cold and never quite stops. Step off that bedrock and the flower is gone. One of the most narrowly pinned plants in the genus.

Same genus. Same Pacific West. Two flowers that have answered the same question, how much world can one kind of orchid live in, in opposite directions. One took nearly a continent. The other took a few wet acres of poison rock. The reason behind the contrast is not the reason you would guess, and a third western slipper complicates it beautifully.

The generalist that stayed loyal

Let me give you the wide one first, in the flesh. The mountain lady's slipper (*Cypripedium montanum*) stands twenty-five to seventy-one centimeters tall (ten to twenty-eight inches), a stem of three to six pleated leaves with, at the top, usually one or two flowers, occasionally three. The pouch is dull white, washed at its base with fine purple veins and spotted deep purple inside, nineteen to thirty-three millimeters long. But look at the petals to either side of the pouch. They are long, narrow, the deep maroon-brown of old leather, and they corkscrew, each one twisting away to its side in one to three full turns, like a dark ribbon pulled taut and spun. That spiral is the field mark you carry away. White pouch. Dark twisted petals. One or two flowers on the stem.

*A mountain lady's slipper (*Cypripedium montanum*), its white, purple-veined pouch set against long, dark, spirally twisted petals.*

It is a handsome plant, and people have said so for a long time. The best line ever written about it belongs to John Muir, who met it in Yosemite and put it in *Our National Parks* in 1901: "*Cypripedium montanum,* the only moccasin flower I have seen in the Park, is a handsome, thoughtful-looking plant living beside cool brooks. The large oval lip is white, delicately veined with purple; the other petals and sepals purple, strap-shaped, and elegantly curved and twisted." Read that again and notice the word he reached for. Not pretty. Not showy. *Thoughtful.* A naturalist looked at a flower and the adjective that came to him was the one you would use for a person sitting quietly with a problem. Twenty-four years later, Mary Vaux Walcott, painting the continent's wild plants for the Smithsonian, came on the same species in bloom in the Columbia River valley: "a wonderful sight burst upon us. Scattered among the low-growing bushes were great clumps of this splendid orchid in the perfection of bloom." She made it Plate 1 of *North American Wild Flowers*, the very first plate of the entire five-volume project. Of all the orchids she could have opened with, she chose this one.

So the mountain lady's slipper has admirers and an empire. And underground, it is a fussy specialist.

You will remember that a lady's slipper cannot germinate alone, that it spends its whole infancy and much of its adult life fed by a fungus it did not choose, and that we settled on calling that dependence not fragility but trust. For most lady's slippers the partner comes from one narrow group of fungi, the Tulasnellaceae, and the mountain lady's slipper is a model of the rule. When Richard Shefferson and his colleagues sampled the roots of twelve mountain lady's slippers in 2005 and asked which fungi lived inside them, the answer came back clean and narrow: Tulasnella, and essentially only Tulasnella. The most far-flung lady's slipper on the continent keeps the strictest possible diet.

That ought to be a paradox. A plant that lives in Alaska and Wyoming and the Klamath and the Saskatchewan plains, on limestone and granite and old volcanic ground, in wet brookside shade and dry open pine, depends on one small family of fungi everywhere it goes? How does the pickiest eater end up with the biggest range? Hold the question; the other two flowers will answer it.

Two more things about this plant. It is one of the two most drought-tolerant lady's slippers we have; the wetland scientists rate it **Facultative Upland**, which is jargon for "you'll generally find it on dry ground." And its bees are a surprise. When Peter Bernhardt's team worked out the pollination at two Oregon sites, they found the mountain lady's slipper is worked mostly by small female sweat bees in the genus *Lasioglossum*, five to ten millimeters of bee, with a few other small solitary bees helping. Bumblebees try and fail; they are too big, sieved out at the door the way the genus does to everything the wrong size. The payoff is remarkable. Between fifty-two and eighty-five percent of flowering stems set fruit, the highest fruit-set recorded for any North American lady's slipper, startling for a flower that feeds its visitors nothing. The likely reason is that small-bee crowd: several redundant species of the right size, abundant on an open conifer floor, all getting fooled.

There is one more story this plant carries, a quieter one. The mountain lady's slipper is, almost certainly, the first lady's slipper any European-American wrote down. On the thirtieth of June, 1805, somewhere above Lolo Creek in what is now Montana, Meriwether Lewis described a flower "in shape and appearance like ours only that the corolla is white, marked with small veigns of pale red." That was this one. The next June, on the way home, Clark noted it again at Weippe Prairie in Idaho. It is the only orchid the Lewis and Clark expedition recorded in its journals. And yet no specimen of it survives in their collection; the surviving herbarium holds no orchid sheet at all. Lewis saw it, judged it not new, and pressed nothing. The flower waited twenty years for David Douglas to collect a type in the Blue Mountains, and fourteen more for John Lindley to publish its name in 1840, six years after Douglas had died on a Hawaiian mountainside. The orchid was written down, then left, then named by men who never stood where it grew.

Here is what I cannot walk past. Weippe Prairie, where Clark noted the flower in 1806, is Nimíipuu homeland, the country of the people we call the Nez Perce, and sits today inside Nez Perce National Historical Park. The Nez Perce gathered camas on that prairie for generations, in the weeks the mountain lady's slipper would have been in bloom around them. The expedition's "discovery" happened on ground that had been known and tended for a very long time. And when you go looking through the published ethnobotanies of the Plateau and the interior Northwest, through Turner and Gunther and Hunn and the rest, for a recorded Nez Perce or Salish or Kootenai use or name for this particular flower, you find nothing tied to *montanum* specifically. That absence may mean the relationship was never written down rather than never existed; the published record is a settler record and a thin one. I can only state the two facts and leave them side by side: the flower's Western scientific origin is sited in Nimíipuu country, and the Western record holds no Nez Perce name for it. The right people to ask are the Nez Perce themselves.

The specialist pinned to one address

Now the opposite flower. If you have only ever pictured a lady's slipper as one nodding pouch on a modest stem, the California lady's slipper (*Cypripedium californicum*) will rearrange your sense of the genus. It is tall, twenty-five to a hundred and twenty centimeters (ten inches to nearly four feet), and leafy past all its relatives: five to ten pleated leaves up the stem, more than any other North American lady's slipper. And it does not hang one flower at the top. It hangs a raceme of them, three to fifteen, as many as eighteen on a strong plant, opening in sequence up the spike over weeks: a single stalk taller than your knee, set with a dozen small white pouches, the lowest already browning while the topmost is still a bud. If you find a lady's slipper with a column of pouches up a leafy stem, you have found this one, and almost certainly found it standing in water.

The pouches themselves are small, the smallest white pouch in North America, only fifteen to twenty millimeters. The petals to the side are pale yellow-green and lie flat and spreading, not twisted at all, which is the close-up way to tell it from the mountain lady's slipper on the next ridge. But the real identifier is underfoot.

*A California lady's slipper (*Cypripedium californicum*): a single white pouch with flat, pale yellow-green sepals. In the wild it carries many such pouches up one tall stalk, in cold serpentine seeps.*

This flower is an obligate of serpentine seeps. **Serpentine** is the surface expression of rock from deep in the earth's mantle, and the soils it makes are heavy in magnesium, starved of calcium, low in the nitrogen and phosphorus and potassium plants run on, and salted with nickel and chromium and cobalt, metals that kill ordinary roots. To most plants, this ground is poison. The water seeping through it comes out cold and alkaline, around pH eight. Most of the regional flora cannot live there, which is exactly why a specialist can. The California lady's slipper grows in these seeps and in the fens kept wet by them, very often in the company of *Darlingtonia californica*, the cobra lily, a pitcher plant that solves the same nutrient-poor ground by eating insects. Find a slope of cobra lilies with cold water running through it, and look for the orchid.

And notice what this habitat does to the plant's freedom. The mountain lady's slipper is rated dry-loving, Facultative Upland; the California lady's slipper is rated **Obligate Wetland**, the far end of the same scale. It cannot leave the water, and it cannot leave the rock, and the rock cannot move. This is the most thoroughly pinned habitat of any lady's slipper on the continent: a serpentine outcrop cannot migrate north as the climate warms, and a seep that dries in August is a seep with no orchid. The flower is bolted to an address that can be neither relocated nor adjusted.

So far this looks like a tidy little fable. Wide flower, narrow diet; narrow flower, you would assume, narrow diet to match. The fungus is where the fable breaks.

When that same 2005 study sampled the roots of the California lady's slipper, it did not find the strict Tulasnella diet that nearly every other lady's slipper keeps. It found the opposite. This one flower associated with three different families of fungi, the tulasnelloid, the ceratobasidioid, and the sebacinoid, in roughly equal share. Of all the lady's slippers in that study, the one bolted to the narrowest possible patch of ground turned out to be the least picky partner of any of them, spread evenly across three whole families where its relatives clung to one. The pinned specialist is, underground, a generalist.

Your first guess about why is probably the same as mine was, and it is wrong. You might think serpentine breeds its own metal-tolerant fungi, that the orchid casts a wide net because it must in such strange soil. But the work has been done, and serpentine soils do not host a distinctive heavy-metal fungal flora; the fungi down there are ordinary woodland fungi. The broad diet is not an adaptation to poison rock. It is just a flower taking whatever competent partner the microsite offers, three families' worth.

And it is the same puzzle the mountain lady's slipper raised from the other direction. Why is the broad-minded generalist the one with the tiny range, and the picky specialist the one that conquered the West?

What a range map is really a map of

Here is the turn, and it is simpler than the puzzle made it look. We have been reading these range maps as if they were maps of the flowers. They are not, quite. The thing that decides where a lady's slipper can live is its fungus, because without the right fungus in the ground a seed cannot even begin. A lady's slipper can only grow where its partner already grows. So a range map of the orchid is, underneath, a map of where its fungus is, fenced in by wherever else the plant needs.

Read the three western slippers through that lens and the paradox dissolves.

The mountain lady's slipper is picky about its fungus, yes, strict Tulasnella only. But its Tulasnella partners are themselves everywhere: common, widespread soil fungi that turn up across the whole conifer West, on limestone and granite alike, north into Alaska and east onto the plains. So the orchid's narrow diet costs it nothing, because the one food it eats is laid out across half a continent. It can live almost anywhere because its partner does. The strict specialist ranges wide not in spite of its pickiness but because the thing it is picky about is itself a generalist of place.

The California lady's slipper is the reverse, and now you can see why. Its broad fungal taste should make it free; three families of partners is a wide menu, available in soils everywhere. The fungus is not what holds it. The address is. It is welded to serpentine bedrock that occurs only in patches, and to seeps that have to stay cold and wet and never fail, and that combination exists in one mountain province in one corner of the West. The orchid could eat almost anywhere. It can stand in only a few places. Its range is small not because its fungi are narrow but because its rock is.

Two flowers, two opposite range maps, one rule under both: the orchid goes where its requirements already go. We spent an earlier chapter learning to read a lady's slipper's dependence on its fungus as a kind of trust. What the western slippers add is that the same dependence, projected outward onto a continent, becomes geography. Pickiness about a widespread partner buys you a whole West. Open-mindedness about your fungus buys you nothing if you have chained yourself to a rock that cannot travel.

One more fact about the California flower shows how strange "how threatened is this plant" can get. Three careful authorities have looked at the same orchid and disagreed about how worried to be. The

International Union for Conservation of Nature rates it Endangered. NatureServe, weighing the same data, calls it Vulnerable, a notch less dire. The California Native Plant Society, with the state agencies, rates it lower still, effectively "apparently secure." The disagreement is honest: a plant pinned to a handful of seeps can look secure if you count its acres as stable and intact, and endangered the instant you remember how few acres there are and how little it would take to ruin one. Both readings are true at once. And the threats are not abstract. In March of 2024 a company called Homeland Nickel staked a hundred and fifteen claims, more than two thousand acres, on Eight Dollar Mountain in Oregon, prime serpentine country, after the nickel locked in that poison rock. Meanwhile a water-borne root disease, *Phytophthora lateralis*, is killing the Port-Orford-cedars that shade these seeps, carried to the trees by the same seep water that feeds the orchid. The flower's whole life runs through that cold trickle, and so, it turns out, does at least one thing that can take it down.

The third strategy, and the orchid the books walked past

There is a third lady's slipper out here, and it does not fit the pair. It breaks nearly every rule of the genus, and it is the strangest of the three. Meet the clustered lady's slipper (*Cypripedium fasciculatum*), also called the brownie. It is the shortest lady's slipper in North America, six to thirty-five centimeters (two and a half to fourteen inches), often no taller than your hand. The leaves give it away first. Where every other native arranges its leaves scattered up the stem or gathered at the base, the brownie carries a single pair set sub-opposite, offset just slightly, spreading wide near the middle of a densely hairy stalk. Above them droops a tight little cluster of usually two to four small flowers, nodding, the source of the name *fasciculatum,* "growing in a bundle." The flowers are small and dull, greenish-yellow stained with reddish brown to dark purple, the pouches among the smallest in the entire genus. The whole plant is built to be missed: short, dark, flowers hung downward, growing in the duff-strewn shade of conifer forests where dim brown things disappear.

The clustered lady's slipper (Cypripedium fasciculatum), *shortest of the slippers, with one pair of spreading leaves and a drooping cluster of dull brown-purple flowers.*

And missed it was. The great popular wildflower book of its own time and place, Mary Elizabeth Parsons's *The Wild Flowers of California,* came out in 1897 and described both of its western cousins, the mountain lady's slipper and the California lady's slipper, in loving detail. It left the brownie out. No Muir passage names it. Walcott never painted it. No state claims it as an emblem, no folk story attaches to it, no moccasin tale, and no nineteenth-century medicine trade ever bothered with a plant this small and this dark. The eastern pink lady's slipper carries centuries of human attention; this one carries almost none. The silence is not an absence of evidence about the flower. It is the truest thing about it. This is the orchid too plain to be loved.

Which would be merely poignant if the brownie were not, underground, doing something more radical than any of its showier relatives. Go back to the rule once more. A young lady's slipper, before it greens up, lives entirely off its fungus. We have always said the adult then turns the corner: it grows leaves, takes up photosynthesis, and lives mostly on its own sugar, keeping the fungus on as a junior partner but no longer fed by it. That is the standard picture of a green lady's slipper. The brownie does not honor it.

In 2005 Darlene Southworth and her student went into three southern Oregon populations and asked two questions of the brownie's roots: which fungi live there, and where is the plant actually getting its carbon. The answers were both off the script. The fungi were not the genus-standard Tulasnella. They were **ectomycorrhizal** forest fungi, the kind that sheathe the root tips of conifers and trade nutrients with the trees: *Russula, Lactarius, Tomentella, Suillus,* the ordinary partners of a Douglas-fir. Out of all those root samples, with primers tuned specifically to catch Tulasnella, they found Tulasnella exactly once. And then the carbon. Using the stable-isotope signatures that betray where a plant's atoms come from, they measured how much of this green, leafy, photosynthesizing orchid's carbon was coming up out of the fungus rather than down out of its own leaves. The number was forty-two to forty-six percent. Nearly half. For nitrogen, ten to eighteen percent. The word for it is **partial mycoheterotrophy**, the half-and-half feeding we first called being a mixotroph, carried further than any green slipper has a right to carry it.

Sit with that. A green plant, standing in the light with functioning leaves, is pulling almost half its carbon from a fungus, and that carbon was fixed not by the orchid but by the surrounding conifers, photosynthesized in Douglas-fir needles, traded into the fungal network underground, and siphoned off by an orchid plugged into the same network. The brownie is feeding partly off the trees, through the fungus, while pretending to make its own living in the sun: a green orchid behaving like one of the ghostly leafless coral-roots that gave up photosynthesis entirely, a flower caught halfway out the door. And this is the only direct measurement of its kind for any North American lady's slipper. Of all twelve, the one we have actually caught in the act of stealing carbon is the little brown one the wildflower books ignored.

I have to be careful here. This single study, the only one ever done on the brownie's fungi specifically, says ectomycorrhizal forest fungi and near-half carbon theft. But a broader, much-cited paper that drew on the very same root samples concluded that lady's slippers as a genus partner "almost exclusively" with Tulasnella, the strict diet we have leaned on all book long. The brownie was the largest single sample in that shared dataset, sixteen plants, and the two peer-reviewed papers read it in opposite directions. Nobody has gone back with modern tools to settle it. So I will not tell you flatly what the brownie's fungus "is." The best species-specific evidence points somewhere genuinely strange, disagrees with the genus rule, and has sat unresolved for twenty years. The one careful look anyone has taken stands crosswise to everything else we know about the genus, and no second look has come to break the tie: the brownie keeps its strangest secret a little longer.

The brownie has one more surprise, and it pays off something promised back when we first took apart the pouch. Every other lady's slipper in North America is pollinated by a bee, or, in one northern exception, fertilizes itself. The brownie alone is pollinated by a wasp. Not just any wasp: a tiny female in the genus *Cinetus*, of the family Diapriidae, still undescribed by science, under five millimeters long, whose larvae are **parasitoids** that grow up by consuming the larvae of fungus gnats. She visits the rewardless flower, gets fooled, and carries its pollen, the only parasitoid-wasp pollination system among all twelve North American lady's slippers, confirmed across populations in Oregon, Colorado, and Idaho. Why she comes is the open question. The leading idea is that the flower smells like the fungus-gnat larvae she hunts as nurseries for her young, a chemical lie that says "lay your eggs here" where there is no host at all. That a lady's slipper can pull off a deception of this shape is not idle speculation; it has been proven in an Asian relative of the brownie, *Cypripedium fargesii*, whose flowers chemically counterfeit a mold the fungus-eating flies they fool come hunting for. The brownie's version, the gnat-larva nursery that lures the wasp, is a different deception aimed at a different insect, and that specific version is the part still unproven. Its odor has been described, qualitatively, as musky and barnyard with a citric edge. But no one has ever run the chemistry, so the mimicry stays a good hypothesis and nothing firmer. And for a flower that offers nothing and is worked by a single small wasp, the brownie sets a surprising amount of fruit, eighteen to sixty-eight percent depending on the population, far above the five to twenty-five percent typical of the bee-deceived eastern slippers (the pink lady's slipper manages only five to thirteen). The full chain, said out loud once: orchid lures wasp, wasp hunts gnat, gnat lives off fungus, fungus may feed the orchid. Four links, and direct evidence for only some.

A diapriid wasp, representative of the odd parasitoid family that pollinates the clustered lady's slipper, beside a fungus gnat of the kind its larvae hunt.

One more thread, and it reaches back a chapter. In the last chapter I mentioned a western mountain slipper that broke the family's leaf template and carried Asian relatives of its own, and left it unnamed. This is that flower. The brownie is the sole New World member of an otherwise wholly East Asian group of lady's slippers, section *Trigonopedia* in the most recent genus-wide genetic work, the section that older treatments set aside for this one species alone, under the name *Enantiopedilum,* which means a single ancestral dispersal carried its line from Asia into the western mountains and left it stranded here, the lone American cousin of a Chinese clan. It makes a neat pair with the ram's-head lady's slipper. The ram's-head broke the genus's flower template, free, spreading sepals where every relative has them fused; the brownie broke the leaf template, that one odd mid-stem pair. Two relicts of old crossings, one strange in its flower and one in its leaves, the only New World members of otherwise-Asian sections, at opposite ends of the same continent.

There is a last tension worth handing you straight. Run the brownie through a climate model, as one 2022 study did for six lady's slippers, and the machine reads optimistic to the point of giddiness: between the nineteen-eighties and the twenty-tens the brownie's modeled band of suitable habitat had already slid some five hundred kilometers to the northeast, its area swelling by more than two thousand percent, the largest such expansion of the six. And then you go out and count actual plants. When one team resurveyed more than two hundred known populations, a third were simply gone; modeled across the whole regional database of some twenty-four hundred populations, the loss ran to roughly forty-five percent, piled up at the low, warm elevations where warming bites first. The map says expanding; the ground says dying. A chapter ago the ram's-head showed the same split, its modeled habitat sliding west while its plants crashed; here it returns sharper and near-reversed, the brownie's suitable climate racing northeast while the flower itself vanishes below. And the brownie cannot follow the map: its seed travels maybe two meters from the parent and cannot grow at all without finding its particular fungal partner already waiting in the new ground. A warming map may open a thousand square miles to the north that the brownie will never reach. For a plant this dependent and this slow, that gap is the whole danger.

Map the brownie and it confirms the rule a third way. It is a montane western endemic scattered in patches across eight states, Washington, Oregon, California, Idaho, Montana, Wyoming, Colorado, and Utah, never anywhere continuous. And it can stand only where the conifer ectomycorrhizal network it taps for half its food already runs; the brownie reaches exactly as far as the forest it feeds on, and not a step past it.

And cultivation cannot rescue it, for the chapter's own reason wearing work clothes. You cannot buy a brownie, because no one can supply what it needs. More than forty species of fungi have been pulled from the roots of a single plant, and three separate populations shared only three species in common. There is no single partner a lab can isolate, multiply, and feed to a seedling; the orchid depends not on a fungus but on a whole shifting committee of them, different from forest to forest. To grow a brownie you would have to grow its forest first.

What I went to see

I have stood in one of those serpentine seeps. The Coast Guard had me stationed down at Petaluma in those years, and on a long weekend I drove north into the Klamath country, where two of these three flowers can occur within a few miles of each other and almost never in the same spot. It is a strange place to find an orchid. The ground is the dull green-gray of the mantle rock it came from, scabbed with lichen, and where the water comes out the cobra lilies stand in their hundreds, hooded and translucent. The water is cold even in July, a thin steady trickle you hear more than see, running for longer than any plant standing in it has been alive.

The California lady's slippers were up along the wettest line of the seep, taller than I expected, each leafy stalk hung with its ladder of small white pouches, the lowest already going brown, the highest still closed. A few feet uphill, off the wet, the ground turned to ordinary forest duff under the conifers, and somewhere up in that dim brown litter, though I did not find one that day, the brownie would be holding its drooping cluster close to the ground, plugged into the roots of the trees, eating half its lunch off the Douglas-firs. And far away, on drier slopes I could not see, the mountain lady's slipper would be standing in its modest ones and twos, picky and triumphant, fed by a fungus that had gone everywhere ahead of it. Three flowers. Three answers.

I crouched at the edge of the wet, where the rock turned dark with water, and looked a while at the nearest stalk of white pouches without touching it. The rock under it would not move a meter in ten thousand years, and neither would the orchid. I left it to its rock and walked back up onto the dry ground, trying to hold both flowers in my head at once, the one that took almost everything and the one that took almost nothing, and what set them apart was down in the dark under my boots the whole way back to the car.

The Far North

While the rest of the continent's lady's slippers are still waiting underground for spring, two of these are already pushing up through the last of the snow. That is not a figure of speech. The United States Forest Service, describing the orchids of Alaska, singles out two species as so well adapted to cold that their shoots come up under the snowpack and finish emerging through it: the spotted lady's slipper and the sparrow's-egg lady's slipper. Picture it. A green shoot, no thicker than a pencil, lifting a crust of late snow off its own back, in country where the growing season is a rumor lasting ten weeks. These are the orchids that live where orchids almost stop. This chapter is about three of them, the boreal and Beringian trio of the far north, and they differ from the nine before them in a way that has nothing to do with how they look. We named them a long time ago. We have barely met them since.

A green shoot lifting a crust of late snow off its own back, in country that thaws for only a few months a year.

One of these three has a pollinator that has never once been watched doing its job, anywhere on earth. Another has a fungal partner that has never been studied at all: not one paper, not one thesis, not one culture dish. The third has lived in our books since the year 1800 and has still never had its North American pollinator or fungus observed. These are not obscure weeds nobody noticed. They are orchids, the most studied plant family on the planet, carrying names a century or two old. The names are old. The knowledge is not. Back when we took the pouch apart, I said that for a few lady's slippers somebody would someday have to go back to them the way Darwin went back to the slipper, and watch. This is the chapter that shows you which ones.

The spotted one, and the snow

Start with the one that reaches farthest. The spotted lady's slipper (*Cypripedium guttatum*) is the only native North American lady's slipper with a true circumboreal range. Every other species in this book is ours, an American plant. This one is not only ours. The same species rings the top of the world: across Russia from Belarus to the Pacific, through Mongolia and Korea and northern Japan and the mountains of China, over into Alaska and the Yukon and the western Northwest Territories. One orchid, two continents, a single living thread tying North America's flora to Asia's.

It is small. Twelve to thirty-five centimeters tall (about five to fourteen inches), often shorter at its northern edge, with two clasping leaves partway up the stem and, as a rule, a single flower. Those two leaves are worth remembering. They are a field mark this orchid shares with the palomino, the second of our three, and the botanists group all three in a modern section they call *Bifolia,* the kinship this chapter is built on. The leaf count is a clue to that kinship, not the test of it: the third species breaks the two-leaf rule, as we will see, and belongs anyway. The pouch is the thing you came for. It is shaped like a little pitcher, white, splashed with bold magenta-purple spots and blotches, the marking that gave the plant its name: *guttatum* means "drop-marked," from the Latin *gutta,* a drop, as if someone had flicked a loaded brush at a white flower. The petals to either side are narrow, spreading, not twisted at all. White flower, bold purple spots, two leaves, the far north. There is nothing else like it on the continent.

*A spotted lady's slipper (*Cypripedium guttatum*), its small white pouch boldly blotched with magenta. It is the only truly circumboreal lady's slipper.*

It lives where you would expect a tough northern plant to live: open spruce and birch woods, slopes and banks, taiga clearings, alpine meadows, the edges of the stunted krummholz forest near treeline. Not bog, not fen; this one likes its feet on the drier side of damp. And not the true Arctic tundra. In North America it reaches the Yukon, the Northwest Territories, and much of Alaska, but it does not cross the Brooks Range. North of that wall of mountains, on the Arctic slope, it simply stops. The cold it can take. The tundra it cannot. In a warming world that turns into something cruel. As for the cold, this is one of the two species the Forest Service names for coming up under the snow, its shoot already moving while the snow is still on it, racing a ten-week season slammed shut at both ends by frost. Up here you get to be early, or you get nothing.

Here is the strange thing about a flower this widespread. We know how the spotted lady's slipper is pollinated, because people have watched it happen. They have just never watched it happen here. In a high meadow in Yunnan, at nearly thirty-five hundred meters (about eleven thousand five hundred feet), Hans Bänziger and his colleagues watched small sweat bees in the genus *Lasioglossum* fall into the slick pouch and climb out past the stigma and anther, dusted with pollen, fooled and unfed, the cold-country version of the kettle trap we took apart earlier. A Korean team watched the same kind of bee do the same thing on a mountain in 2022. But every one of those observations was made in Asia. No published account exists of what pollinates the spotted lady's slipper in North America, on the Alaskan and Yukon ground where it has lived all along. The oddity is the breadth. The most far-flung lady's slipper on earth, named for two centuries, and on its own continent no one has ever seen the bee. It is almost certainly a sweat bee here too. But almost certainly is not watched, and nobody has watched.

The underground partner is darker still. A lady's slipper is nothing without its fungus; the seed cannot even germinate until the right fungus finds it and feeds it through years of darkness. For most lady's slippers that partner comes from one narrow group, the Tulasnellaceae, and the best guess is that the spotted lady's slipper is no exception. But it is a guess. The handful of times anyone sequenced the fungus inside this species, the plants were Asian. No North American spotted lady's slipper has ever had its fungus sequenced. The orchid connects two continents in a single species, and we have looked at the roots on only one of them.

What the name does not tell you

We have had the name *Cypripedium guttatum* since the year 1800, when the Swedish botanist Olof Swartz published it from material that had traveled out of Siberia. And the plant had been drawn eighty years before that: a German naturalist, Messerschmidt, on a Russian expedition into Siberia, sketched this spotted flower around 1720, a full lifetime before it carried a Latin name. By the time the Swedish botanist Eric Hultén spent the early twentieth century mapping the floras of Kamchatka and the Aleutians and Alaska, the spotted lady's slipper had been a known, named, illustrated species for generations. And in all that time, on this continent, no one has watched it be pollinated or learned what feeds it underground.

That gap holds for all three orchids in this chapter. Naming is not knowing. To have named a flower is not to have watched a bee crawl out of it, or to know the fungus that nursed it from a speck, or to have stood in the right meadow in the right ten days and seen it live. We assume that because a thing has a name in a book, somebody did the watching. For these three, mostly, nobody did. The label went on first and the life behind it stayed unread. The far north is full of orchids we have been on a first-name basis with for a century and have never actually met.

The grouping itself proves the point. Placing these orchids in section *Bifolia* is a recent decision: it rests on a 2024 study by Lagou, Kadereit, and Morales-Briones that read nearly a thousand stretches of their DNA. Older botanists had filed the sparrow's-egg somewhere else entirely, in a section called *Obtusipetala*, next to the showy lady's slipper we met in the ninth chapter. And the 2024 work confirmed only two of the three by name, the spotted and the palomino. Whether the sparrow's-egg truly belongs in *Bifolia* with them, the molecules have not yet said. So even the family tree of this trio is a half-answered question. We sorted them into a section before the data could tell us the section was right.

Spotted everywhere, vanishing somewhere

There is a second lesson in this flower, about the comfort we take from conservation rankings. By the global numbers, the spotted lady's slipper is one of the most secure orchids in the world: NatureServe rates it G5, "secure," and the International Union for Conservation of Nature lists it as Least Concern. The labels are not wrong. A species in thousands of populations from Belarus to Alaska is in no danger of disappearing, and global redundancy is all a global rank measures. It says nothing about any particular edge. And at the edges, the same secure orchid is in free fall. In South Korea the spotted lady's slipper is Critically Endangered, down to two known populations on a single mountain, dug and sold by people who want it for its beauty; in Japan it is Critically Endangered as well, down to a single northern-Honshu population of about a hundred plants. The flower an American database calls "secure" is, on the far side of its range, being poached to the edge of national extinction. A wide range can hide a local catastrophe completely; "secure" is an average, and an average can stay comfortable while its extremes come apart.

Climate adds a turn. When a team modeled how the suitable ground for six North American lady's slippers had shifted between the nineteen-eighties and the twenty-tens, five slid generally northward, as a warming-pressed species should. The spotted lady's slipper did the opposite. Its suitable ground shifted *south*, and uphill. The reason is the Brooks Range, and the ocean past it: this is a flower already pinned against the northern limit of the landmass, blocked from the Arctic tundra it cannot tolerate, so when the climate warms under it, there is no north left to retreat into, only up. A plant the maps call secure is, at its own northern edge, climbing for its life with the sea at its back.

Two smaller truths round out the portrait. Of all twelve native lady's slippers, the spotted one is the cleanest case of a flower that never went through the old root trade; its range lay outside the Appalachian and Great Lakes country where nineteenth-century gatherers tore lady's slippers from the ground by the ton for the medicine market, and it was too far north to be worth anyone's shovel. There are scattered reports of other uses far away, a Chinese stem soup, a Mongolian and Tibetan heart remedy, but they come down different lineages and both reach us secondhand and unverified, and I will not turn a thinly sourced rumor into a recipe. And the human record that should sit closest to this plant is empty. The annotated bibliography of Aleut and Unangax̂ ethnobotany compiled under the Conservation of Arctic Flora and Fauna program documents four other orchids, each attributed to named knowledge-holders, but contains no *Cypripedium* of any kind. The standard index of Ainu medicinal plants, covering the homeland that overlaps this orchid's range in northern Japan, holds one orchid and no lady's slipper. That is a documented gap in the published record, not a claim that the relationship never existed. The record is a settler record and a partial one, and a silence in it may reflect what was never written down rather than what was never known. The right people to ask are the descendants of those named knowledge-holders, and that asking has not happened.

The palomino, and the swarm nobody has sorted

Now the rarest of the three, and the least known orchid in this entire book. Out on the Aleutian chain, that long fog-bound arc of treeless volcanic islands curving away from Alaska toward Kamchatka, there grows a lady's slipper that most people, including most botanists, have probably never correctly identified. Meet the palomino lady's slipper (*Cypripedium yatabeanum*). I use "palomino" on purpose, the name the native-orchid people prefer, because the other English name in the books, "spotted lady's slipper," collides with the species we just left. The collision is no accident: the two are close kin, they meet on these islands, and telling them apart is exactly the problem.

Look at the flower and the palomino name explains itself. The pouch is not the bold magenta-on-white of its spotted cousin. It is cream to pale yellow, flecked and muddied with brown and tan and an olive-tinged maroon, the dull dappled coloring of a palomino horse's coat, and longer and narrower, a tapering sac without the little out-turned pitcher mouth the spotted one has. And the petals are the giveaway under a hand lens: where the spotted lady's slipper has plain narrow petals, the palomino's are pinched in near the middle, fiddle-shaped, what a botanist calls **pandurate**. Two clasping leaves, the same family signature. Seventeen to thirty-two centimeters tall (about seven to thirteen inches).

*A palomino lady's slipper (*Cypripedium yatabeanum*), its dappled cream-and-brown pouch, photographed in Japan. In North America it clings only to the Aleutians, and almost no scientist has studied it.*

This is the most range-restricted native lady's slipper in the United States, by a wide margin. On the whole continent it grows only on the Aleutian Islands, the Alaska Peninsula, and Kodiak Island. That is it. Everywhere else it lives is across the water in Asia: Kamchatka, the Kurils, Sakhalin, Hokkaido, and the high mountains of central Japan. North America gets the last toehold of an Asian flower that leapt the Bering gap and got no farther.

On the islands where the palomino meets the spotted lady's slipper, the two interbreed, and the cross has a name of its own, *Cypripedium ×alaskanum*. The hybrids do not stay rare. They take over. The botanist Charles Sheviak, writing the authoritative North American treatment, put it about as bluntly as a flora ever does: most reports and illustrations of the palomino lady's slipper in Alaska are, in fact, based on hybrids. A field worker who went out to one Aleutian island in recent years found hundreds of hybrids and only a couple of pure plants. The orchid you are looking at is more likely than not a blend, and almost nobody has done the patient work of sorting which plants are the genuine article. And it gets less settled the closer you look. When a 2024 study brought nearly a thousand stretches of DNA to bear on the genus, it could not even confirm that *×alaskanum* is the straightforward hybrid we have long called it, and in the same breath raised the possibility that the spotted lady's slipper itself is a hybrid lineage, an ancient blend, not a clean species at all.

Against that backdrop, the list of what we do not know about the pure palomino is staggering. Its pollinator has never been observed. Not in Alaska, not in Japan, not in Russia, not anywhere, ever. We infer a sweat bee, because the hybrid exists and a hybrid needs a shared pollinator and the cousin uses sweat bees, but inference is all we have. Its fungus has never been sequenced, in any country. There is no isotope study of how it feeds, no protocol for growing it from seed, no long-term count of its populations. Every line of inquiry you could open on a plant is, for this one, still closed. Of all twelve native lady's slippers, this is the one science has looked at least, the genus's canonical "we don't know."

The few things we do know are about the world the palomino lives in rather than the flower itself. That world is stark. No trees. Fog off a cold sea, wet volcanic ground underfoot, the air loud and rank with nesting seabirds, and somewhere in that meadow this pale dappled pouch on its short stem. Those meadows are fertilized, measurably, by the sea: vast seabird colonies nest on these islands, and their droppings carry ocean nitrogen onto the land and feed the whole green system, meadow and orchid alike. When fur traders introduced Arctic foxes to farm them for pelts, the foxes ate the seabirds, the guano subsidy collapsed, and the lush meadows thinned. A flower like this is a quiet readout of that chain, of whether the birds are still nesting and the sea still reaching inland through them. Far away, in the alpine zone of the Japanese mountains, the only directly documented threat is more mundane, overabundant sika deer browsing it down.

And then there is the name, the one rich, fully documented thing about this otherwise undocumented orchid. The palomino is named *yatabeanum* for Ryōkichi Yatabe, who collected the original specimen on a Japanese mountain in 1884. Yatabe was the first professor of botany at Tokyo Imperial University and the founding director of its botanical garden, and a poet besides, co-author of the foundational 1882 anthology that launched modern free-verse poetry in Japanese: a botanist-poet at the hinge between Western science and a new Japanese literature. The name was published in 1899 by Tomitarō Makino, Yatabe's former protégé, in the very journal Yatabe had co-founded. The two men had fallen out years before; Yatabe had banned Makino from the university's botanical institute. Then Yatabe drowned. In the summer of 1899, within months of his death, Makino published this orchid under his estranged mentor's name, a memorial laid down by the man he had quarreled with. An American island chain points west across the Pacific, and at its far tip blooms the name of a Japanese botanist-poet, given by someone making peace too late.

Ryōkichi Yatabe (1851–1899), the Japanese botanist and poet for whom the palomino lady's slipper is named. Public domain.

Its Japanese vernacular names carry their own freight: the "yellow Atsumori plant" and the "small Kumagai plant," after the medieval war epic *The Tale of the Heike*, the slain boy-warrior Atsumori and Kumagai, who killed him and became a monk in grief. It is, too, the one lady's slipper in this book whose Russian name keeps the man himself; in the Russian Far East it is *bashmachok Yatabe*, Yatabe's slipper, the eponym carried into a third language rather than translated away. As for the Indigenous record of the peoples whose islands this orchid grows on, it is a documented blank: the same Aleut and Unangax̂ bibliography that named four other orchids holds nothing for this one, and the Kodiak and Ainu records that should sit nearest it hold nothing either.

The sparrow's egg

The northernmost lady's slipper on the continent is the one I find hardest to leave. The sparrow's-egg lady's slipper (*Cypripedium passerinum*) grows farther north than any other in this book, straight across the boreal and sub-Arctic from Alaska through all three Canadian territories and down the spine of the western provinces, reaching its lonely southern outpost in the mountains of northern Montana. Twelve to thirty-eight centimeters tall (about five to fifteen inches), usually a single flower. And here is the field mark that separates it at a glance from its two cousins: where the spotted and palomino lady's slippers each carry two leaves, this one carries three to five, sometimes as many as seven, stepping up the stem. Count the leaves. You have your answer. This is the species that breaks the two-leaf rule, the *Bifolia* member that does not keep the count, and it is also the one whose place in the section the 2024 molecules left open.

The pouch is the reason for everything, including both its names. It is a small, inflated, nearly pure white sac, eleven to twenty millimeters long, and the white is so thin that the faint crimson and purple spots on the inner wall show through it from inside, like markings seen through eggshell. That is precisely what its first describer saw. And the lower two sepals behind the pouch are usually fused into a single piece, a synsepal: the structural opposite of the ram's-head lady's slipper we met two chapters back, whose lower sepals stay always free and spreading, the two little horns. Same family, opposite solution to the same piece of anatomy.

*A sparrow's-egg lady's slipper (*Cypripedium passerinum*), its small, nearly closed white pouch finely spotted within, like markings seen through an eggshell.*

This is the second of the two species the Forest Service names for coming up under the snow. It lives at the cold edge of the world, in moist mossy spruce woods and, very characteristically, in the coarse soils of glacial river terraces and streambanks, the kind of fresh ground that floods and shifts. That preference matters more than it looks, and it connects to the strangest thing about the flower.

You met this orchid once before, in the sixth chapter. It was the rule-proving exception: the one North American lady's slipper that mostly skips bees and fertilizes itself, autogamous, the best-documented self-pollinator in the genus on this continent, its pollen turning fluid a few days after the flower opens and sliding down onto its own stigma under gravity, no insect required. Paul Catling and others read that habit as a colonist's tool. For a plant making its living on ground deglaciated less than ten thousand years ago, self-pollination is fertility insurance, the trait that lets a pioneer move onto fresh land and not wait on a bee that may never come.

There is one more wrinkle, and it is the best in the genus. The sparrow's-egg does not self-pollinate everywhere. In the unglaciated corner of southwestern Yukon, a patch of the far north the ice sheets never covered, part of old Beringia, there lives a different kind of sparrow's-egg, an outbreeding form, its anther held away from its stigma so self-pollination cannot happen by gravity. When Catling and Bruce Bennett examined those Beringian flowers, self-pollination had failed in seven of the eight they checked. Everywhere the ice scoured the land bare and the orchid had to recolonize, it became a selfer, because a selfer can pioneer. In the one ancient refuge the ice never touched, it kept the old outbreeding way. The exception has its own exception, sited exactly where the logic predicts. This is the live edge of the biology, the part still being read.

And underneath the sparrow's-egg lies the emptiest fungal record in the entire genus.

I have told you, species after species, that we are thin on the fungal partners of these orchids. For the sparrow's-egg it is not thin. It is nothing. No one has ever sequenced its fungus, cultured it, looked at it under a microscope, or written a thesis on it. When the one genus-wide survey went looking at lady's slipper fungi across the world, the sparrow's-egg was not among the species sampled, and it has appeared in no fungal study since. The ram's-head lady's slipper, the data-poorest of the others, at least has a single thesis to its name. The sparrow's-egg does not have even that. Every other lady's slipper in this book has at least a sliver; this one has a clean blank where the answer should be. The northernmost lady's slipper on the continent, the boldest pioneer of fresh ice-freed ground, the strongest self-pollinator we have, and we do not know the name of the fungus that feeds it. Nobody has looked.

I find that thrilling rather than sad, because the question almost asks itself: does a flower that pollinates itself, and colonizes recently deglaciated ground, run a narrower fungal partnership than its outcrossing relatives, or a broader one? You can argue it either way, and there is a natural experiment built right in, the selfing populations on deglaciated ground against the outbreeding relict in old Beringia. The blank is not a dead end. It is an open door with a draft coming through it.

A name carried out of a catastrophe

The sparrow's-egg has the richest naming story of the three, and it shows both halves of this chapter at once: how completely we can document the name of a thing while knowing so little of its life.

The flower was first found by Thomas Nuttall, the great wandering naturalist, who collected it but published no description and meant to call it *Cypripedium album*, a name already taken. It fell instead to John Richardson to name it, in 1823. And the circumstances are extraordinary. Richardson was the surgeon and naturalist on Sir John Franklin's first overland expedition into the Canadian Arctic, the Coppermine expedition of 1819 to 1822, one of the most catastrophic journeys in the history of exploration: of the roughly twenty in the overland party, eleven died on the starvation retreat, reduced at the end to eating lichen and the boiled leather of their boots. And out of that nightmare, among the specimens that came home, was a small white orchid Richardson described for the botanical appendix to Franklin's published narrative.

What Richardson wrote of the pouch is the documentary heart of this flower, and I will give it to you in his own Latin, because the whole identity of the species comes from this one phrase. He called the lip *"magnitudine vix ovi passeris"* — hardly the size of a sparrow's egg. That is the source of everything. The Latin epithet *passerinum*, "of the sparrow," comes straight out of *passeris* in that line, and the English name, sparrow's-egg lady's slipper, is a plain translation of it. A surgeon home from an expedition that killed more than half its men looked at a small white spotted pouch and reached not for a measurement in millimeters but for the egg of a common little bird, and that homely, exact image has carried the flower's name for two hundred years: the whole identity of a species, fixed by one man's eye for the right comparison, on the far side of starvation.

A single sparrow's egg, at scale. John Richardson named the flower for a pouch he thought hardly the size of one.

There is a second naming layer, gentler: in 1929 the Québec botanist-monk Brother Marie-Victorin described a variety of the sparrow's-egg from the Mingan Archipelago, the limestone islands along the north shore of the Gulf of St. Lawrence, where the flower makes its confirmed eastern stronghold. So it carries two human stories at its edges, a British Arctic catastrophe at one end of the range and a French-Canadian friar's careful island botany at the other, both written down in full while the living flower kept its underground secret.

The conservation picture has the same shape as the spotted lady's slipper's, so I will be brief. By the global rank the sparrow's-egg is secure, G5, safe across its enormous boreal range; the international assessors, weighing a different axis, call it Vulnerable, because the area it actually occupies is small and declining. Neither is lying; they measure different things, breadth versus density. The cleanest documented decline is from Pukaskwa National Park on the Ontario shore of Lake Superior, the southeastern edge of the range, where a long study found the orchid simply failing to recruit new plants. Like the spotted lady's slipper it escaped the medicine trade by sheer remoteness, even more invisible than its boreal cousins, missing entirely from the great nineteenth-century catalog of useful plants. And the Indigenous record is, once more, a documented absence: the named-source ethnobotanies of the Gwich'in and the Cree and Dene and Métis were searched, and none carries a properly attributed name or use for the sparrow's-egg specifically, even though those same sources speak in detail about other lady's slippers. The silence sits in the record, not in the knowledge.

What is still out there

I have not seen any of these three flowers, and I will not say I have. The palomino lives on fog-wrapped islands at the end of a thousand-mile chain, in a hybrid swarm hardly anyone has sorted; the spotted one rings the top of the world but keeps to country I have never reached; the sparrow's-egg grows north of every road I have driven. I have only ever seen them the way you have, in photographs and in the hedged sentences of the few who got out there. For me, as for most of us, the far north's orchids are still rumors with names. I would rather end on that honestly than pretend a boot-print I never made.

So picture the one I would most want to stand in front of. Somewhere on a gravel river terrace in the boreal north, right now, the spruce coming down dark to the water and the new gravel still raw from the last flood, a sparrow's-egg lady's slipper is in bloom: small and easy to miss, three or four leaves stepping up a thin stem, a single white pouch nodding at the top, the faint crimson freckles just showing through the wall when the light comes from behind.

It has very likely already fertilized itself, without help from any bee, days before anyone could arrive. And under it, threaded into its roots, is a fungus that has never been named, doing the oldest work in this book, and no one on earth could yet tell you what it is. Nobody is there to dig it. You could not learn it that way, and you would only kill the plant. It stands on its raw new ground, white pouch tipped toward the river, the most interesting thing about it still down in the dark, still waiting, the way it has waited two hundred years, for someone to come back and finally watch.

Part Four — Learning to Look

To notice one of these flowers, it turns out, is already to begin keeping it. This last part is about where they grow and why they are vanishing, why a wild one cannot be dug up and carried home, and how to find them without loving them to death.

Where They Grow and Why They're Vanishing

There is a trick the best field botanists do that looks, the first time you see it, like a kind of cheating. You walk a stretch of ground with one of them, looking for a lady's slipper, and before either of you has spotted a single flower, she stops, glances around, and tells you whether one will be here. Not by seeing it. By reading everything around where it would be. She reads the rock, whether it is acid sandstone or sweet limestone. She reads the water, where it sits, where it seeps, whether the ground squelches or drains. She reads the light coming through the trees and the trees themselves, pine or cedar or oak. And from those three or four facts she will say, with the quiet confidence of someone reading a sentence she has read a thousand times, that it will be there, on the rise where the seep crosses the slope, or nowhere on the hillside. Then you walk to the rise, and there it is.

I used to think this was intuition, the unteachable thing that comes only from forty years on your knees in the woods. It is partly that. But mostly it is something stranger and more hopeful. The orchid is legible. Each of these twelve plants reads a precise set of conditions, the way a thermometer reads heat, and if you learn to read the same conditions you can find the plant before the plant finds the light. John Burroughs put it in a single sentence in 1871, in *Wake-Robin*, and I have never found a better one: "Show a botanist a landscape, and he will tell you where to look for the lady's-slipper, the columbine, or the harebell." That is the whole of this chapter's first half. Where these orchids grow is not random, and it is not a secret. It is written in the ground.

And the second half follows from the first. Because if a lady's slipper is a precise reading of the ground, then it disappears the moment the ground stops saying what it used to say. These plants almost never vanish to a shovel. They vanish when one condition, quietly, stops being true.

Where they grow: the orchid as a reading of the ground

A north-woods clearing in early summer. Reading ground like this, its rock, water, light, and disturbance, is how you learn where lady's slippers will be.

Start with the rock. The twelve native lady's slippers sort themselves, more sharply than almost any other group of plants I know, along a single axis: the sourness or sweetness of the soil. At one end sit the **acidophiles**, the acid-ground plants, and the genus's signature acidophile is the pink lady's slipper (*Cypripedium acaule*), the one most of eastern North America simply calls the moccasin flower. Stand on a dry Appalachian pine ridge in May, the soil a thin acid skin over sandstone, the duff so slow to rot that your boot sinks an inch into a century of unfinished pine litter, and you are standing where it lives. It holds to soil so acid that most plants sulk in it, down around pH four to four and a half, in pine and pine-hemlock woods, on droughty oak-heath ridges, along the sandy margins of bogs. I told its story earlier, the orchid that took the sour ground nobody wanted and made a home there. The acid is probably no accident. The likeliest reason, though no one has yet caught the mechanism in the act, is that below about pH five the soil microbes that would otherwise overwhelm its narrow fungal partnership are themselves held back, and the orchid can keep faith with the one fungus it needs. The far southern giant, the Kentucky lady's slipper (*C. kentuckiense*), is a different sort of case, and a useful warning against reading the rock too cleanly. It is best understood not by its pH but by its water: a seepage plant of Southern bottomland-hardwood forest, holding to springhead seeps and stream terraces in the Ouachitas and the Gulf woods. The ground it stands on is usually acidic sand, but not always; in parts of its range the same plant sits on calcareous limestone, so what truly binds it is the seep, not the sourness.

At the other end of the axis sit the **calciphiles**, the lime-lovers, plants that need base-sweet, calcium-rich ground and will not grow without it. The small white lady's slipper (*C. candidum*), the candle-white prairie orchid, is one: an obligate of lime-rich, never-plowed tallgrass prairie and prairie fen. The showy lady's slipper (*C. reginae*), the queen, is the most strongly lime-loving of all of them, a plant of calcareous fens and white-cedar swamps where the water comes up through limestone and runs sweet. Between the two poles, leaning toward the lime-and-neutral side, sits the yellow lady's slipper (*C. parviflorum*), the workhorse of the genus across the continent. One orchid took the acid; another took the lime; and if you know which ground you are standing on, you have already ruled out half the genus. The sorting is real, and it is most of the map.

Then there are the extremes, the plants that read something even more specific than acid or lime. The California lady's slipper (*C. californicum*) is bolted to serpentine seeps, the cold alkaline water that runs off rock from deep in the earth's mantle, metal-rich and close to poison for most green things. It cannot leave the water; on the wetland scientists' dryness scale it is an Obligate Wetland plant, pinned more thoroughly to its habitat than any other lady's slipper. The ram's-head lady's slipper (*C. arietinum*), the strange one with the little down-pointed pouch netted in red, is not bound by lime at all but by cold: a cool root zone is the thing it cannot do without, which lets it live in places as unlike as a wet cedar swamp and a dry jack-pine sand flat, so long as both stay cold. And up in the boreal north live three more that puzzle me, because they grow on ground that looks nothing alike yet seem to be reading the same thing. The spotted lady's slipper (*C. guttatum*), the sparrow's-egg (*C. passerinum*), the palomino (*C. yatabeanum*): scattered across variable northern country, what they appear to hold in common is not the soil under them but the season over them, the brevity and chill of a far-north summer. The two western generalists, the mountain lady's slipper (*C. montanum*) and the small dull brownie (*C. fasciculatum*), spread more loosely across montane forest, but even they hold to their own envelopes of shade and damp.

Lay a second axis across the first and the picture sharpens again: water. At one end, the wetland-bound plants of the seep and the fen. The California lady's slipper is the absolute case, the one that literally cannot live out of the water; the queen of the calcareous fens keeps her own company close beside it, a creature of wet sweet ground even if she will, now and then, stand in a merely damp meadow. At the other end, the facultative-upland plants of generally dry ground, the mountain lady's slipper and the pink on its droughty ridges. A lady's slipper is partly a reading of how wet the ground stays in summer.

And a third axis, the one people miss most often: light, and what keeps the light coming. Several of these plants live in landscapes that depend on disturbance even though the individual plant cannot survive the disturbance itself. The small white lady's slipper is the clearest case. It is a prairie plant, and prairie is made and kept by fire; without periodic burning the grassland fills with brush and trees and shades the orchid out. Yet a fire passing over an individual plant in the wrong season will kill it. The plant needs the fire to have come, and needs not to be standing in it. Forests do a quieter version of the same thing: let the canopy close and the brownie and the mountain lady's slipper slide toward darkness and disappear. Light is a condition like any other, and something has to keep renewing it.

Here the first half of the chapter turns practical, and turns into Part Four's project of learning to look. Because each orchid reads such a precise set of conditions, the conditions tell you where the orchid will be. Read the rock, the water, the light, and the trees, and you have read the plant before you have seen it. This is what Burroughs meant. It is also why I keep returning to a fact I laid out earlier and will only touch here: a range map of one of these orchids is really a map of where its fungus grows, fenced in by whatever else the plant needs. The flower is just the part of that reading you can see.

I learned to read these axes, what little I can read of them, on my own ground in West Virginia, which is a good place to learn, because the last ice age left the high country here a scatter of cold northern islands: red spruce, peat bog, open heath. The lady's slippers sort themselves across them in plain sight. On the acid heath of Dolly Sods, on sandstone gone sour, the pink one turns up among the blueberry and the laurel. A few ridges east, where the water rises through limestone in the fens around Canaan, the showy lady's slipper survives at one of its last sites in the state. And the most orchid-famous ground of all, the Cranberry Glades, is a cold sphagnum bog carrying something like sixteen kinds of orchid. Not one is a slipper. No slipper ever found the exact ground it needed there.

A lady's slipper is not really a single plant standing in a place. It is a *standing agreement among conditions,* a settlement holding for now between the soil chemistry and the water table, the light and the disturbance regime that renews it, the fungus underground and the bee in the air. Every one of those is a clause, and the plant is what the agreement produces while all the clauses hold at once. Take that seriously and the way these orchids vanish stops being a mystery.

Why they're vanishing: when a clause lapses

A lady's slipper rarely dies because something dramatic happens to it. It dies because one term of the agreement quietly lapses. The fire that used to come stops coming. The water table drops a foot. The canopy closes over. The prairie is turned under. Even, in one strange case we will get to, the air gets cleaner. The plant is still standing when the condition fails, sometimes for years, living on the reserves in its rhizome, the savings account I described earlier that never refills. Then the account runs dry and the plant is simply gone, and from the outside it can look as though nothing happened at all. Nothing did happen, in the dramatic sense. Something quietly stopped being true. That is almost always the real story, and it is why the threat is so easy to miss until it is finished.

Habitat loss, the quiet dominant

By a wide margin the biggest reason these orchids are vanishing is not the shovel and not the climate. It is the slow failure of one of those conditions across a whole landscape. The name we have for it, habitat loss, makes it sound like a single visible event. Usually it is nothing of the kind.

The clearest case in the whole genus is the small white lady's slipper and the American tallgrass prairie. Something like ninety-five percent of that prairie is gone, broken to farmland in a couple of human generations, and the orchid went down with it. It needed no poacher and no drought. It needed prairie, and the prairie became corn. The same arithmetic, in slower motion, plays out in the fens and seeps the calciphiles depend on: drain the wetland, drop the water table, and the queen and the small white lady's slipper lose the sweet wet ground that was holding them. For the western forest orchids the term that lapses is light and soil. For the brownie, the two threats NatureServe ranks highest are canopy removal from logging and thinning, and disturbance of the soil itself, the duff and the fungal network just under it: take the trees and you take the shade, churn the ground and you cut the fungus the orchid feeds through.

Intact prairie meeting a plowed field at a hard line. Across the tallgrass region this conversion erased roughly ninety-five percent of the prairie, and the small white lady's slipper with it.

And then there is the paradox of fire. So many of these landscapes are fire-dependent (the prairie that the small white lady's slipper needs, the open grassland and woodland edge that several others want) that suppressing fire, keeping every flame out for a century in the name of protection, closes the canopy or chokes the grassland with brush and quietly erases the orchid. The plant needs the disturbance regime to keep running even though it cannot stand inside a single event of it. Stop the fire to save the woods, and you can lose the flower that the fire was keeping room for.

Now and then the loss does arrive as a bulldozer, and those cases are worth naming as the exceptions that prove how rare the dramatic version is. In Knox County, Kentucky, a single development of a big-box store erased a population of the Kentucky lady's slipper, more than five hundred plants at once, a meaningful slice of a species that may number only around three thousand individuals on earth. In Brandon, Manitoba, a proposed housing development sitting on a small white lady's slipper population pushed the matter all the way to a federal imminent-threat assessment in 2023, the rare machinery a government reaches for when local protection is failing. But for every orchid lost to a visible blade, far more are lost to a water table that sank, a fire that never came, a canopy that closed while no one was watching.

I quoted Susan Fenimore Cooper once already, near the start of this book, giddy over a haul of eighteen pink moccasin flowers gathered in a single Cooperstown afternoon in 1850. This is the same woman, the same *Rural Hours,* the same Cooperstown spring, and now the other half of the picture, the line I promised we would come back to. Walking that same country, she had also set down what was already disappearing from ground where it had once been common: "The strange pitcher-plant is said to have been much more common, and the moccasin flower abounded formerly even within the present limits of the village. Both are now rare, and it is considered a piece of good luck to find them." That is, as far as I can find, the earliest clear written record of a lady's slipper's decline in North America, set down before the Civil War, before the great medicinal-root trade had even peaked. She was watching a clause lapse in real time, and she knew it.

The American Valerian trade, and a debt that was never paid

One chapter of human pressure on these plants deserves its full telling, because it is the one most people have never heard and the one that reaches back furthest. For roughly a century, the yellow lady's slipper and, to a lesser degree, the pink were dug from the wild by the ton and sold as medicine.

It began, in the formal American record, in 1828, when the eccentric naturalist Constantine Rafinesque, in his *Medical Flora,* gave the yellow lady's slipper the trade name it would carry for a hundred years: "American Valerian," the homegrown substitute for the European sedative root. The name stuck. The root went officinal, listed as an official drug in the United States Pharmacopoeia, from 1863 to 1916, half a century of formal medical standing. Two kinds of enterprise did the manufacturing. The first was the Shaker pharmaceutical communities, the celibate religious societies at Mount Lebanon in New York and Sabbathday Lake in Maine, whose herb departments processed lady's slipper root at something like twenty-eight to forty cents a pound, in volumes that for a single community could reach two hundred pounds in a month. The second, larger and later, was Lloyd Brothers of Cincinnati, the great wholesale house of Eclectic medicine, founded in 1885, which stocked the root among its hundreds of standardized preparations.

Rafinesque's 1828 Medical Flora, *which listed the yellow lady's slipper as "American Valerian." The naming helped set off a medicinal-root trade. Public domain.*

Look at the scale at the bottom of this trade, out where the actual digging happened. The southern Appalachians were the supply heartland. By the eighteen-eighties, one dealer in western North Carolina was reported to employ something on the order of forty thousand root-gatherers, people fanning out across the mountains to dig medicinal plants, the lady's slipper among them, and by the turn of the twentieth century the federal estimate was that Appalachia alone supplied roughly three-quarters of all the native medicinal plants sold commercially in the United States. Forty thousand pairs of hands, digging. A plant that takes the better part of a decade to reach its first flower and cannot be transplanted alive would simply melt out of the accessible woods.

And here is the part that Part Four's whole ethic of keeping requires us to hold honestly. That trade was settler commerce built on Indigenous medicinal knowledge: knowledge taken first, and then drawn on, largely without compensation and largely without credit. Rafinesque himself, naming the plant for the American pharmacy in 1828, acknowledged it was already long known to the Native peoples whose homelands the diggers worked. The debt that implies was never paid. I am a settler writing this, learning it from the published record rather than from within any of these traditions, and the honest thing is to name the Nations and the sources and then to stop, because the medicinal knowledge itself is not mine to retail. I will give no uses, no preparations, no how-to of any kind.

What I can do is point to where the record runs. The use of these plants is documented for the Cherokee, by James Mooney and Frans Olbrechts in *The Swimmer Manuscript* of 1932 and again by Hamel and Chiltoskey in their 1975 work on Cherokee plants. It is documented for the Haudenosaunee, the people of the Iroquois confederacy that includes the Onondaga, toward whose country Rafinesque pointed directly, in James Herrick's *Iroquois Medical Botany*. It runs through Huron H. Smith's ethnobotany of 1923 for the Menominee, and through M. J. Black's 1980 *Algonquin Ethnobotany* for the Anishinaabe and the related Algonquin. For the Mi'kmaq and the Penobscot, the Wabanaki peoples of the northeast, it reaches the literature through Daniel Moerman's compilations. And the very name most of the East uses for the pink lady's slipper, moccasin flower, is an English calque of the Anishinaabe *makizini-waabigwan*, the moccasin flower, a name I have returned to at the pink and the yellows and return to once more here only to mark whose word it was first. One gap I want to name rather than paper over: among these Nations, the Wabanaki are the one for whom this book does not yet have a living, contemporary corroborator to review the attribution, and that is a hole in the work, not a thing resolved.

The trade ended for three converging reasons. In 1910 the Flexner Report, the survey that reorganized American medicine around laboratory science, condemned the Eclectic and homeopathic schools and pulled the ground out from under the doctors who prescribed the root; the Eclectic Medical Institute in Cincinnati finally closed in 1942. In 1916 the lady's slipper came out of the Pharmacopoeia. And in 1975 the entire genus *Cypripedium* went onto Appendix II of the international wildlife-trade convention, closing wild export, while the herbal-products industry's own trade association moved to prohibit wild harvest of these orchids among its members.

What the trade did to the plants needs stating carefully, because the popular version overstates it. You will read that lady's slippers were "harvested to near extinction," and for the species as a whole that is simply not true: the pink lady's slipper is ranked globally secure to this day. The honest framing, the one the record supports, is that the root trade was the documented proximate cause of severe local depletion of the yellow and the pink lady's slippers across much of their range through the early twentieth century. Severe, local, and real, but not range-wide annihilation.

There is one more thing the trade reveals, and it is quietly beautiful as a piece of natural history. The trade was thick for exactly two species, the yellow and the pink, and thin to nonexistent for the other ten, and each of the ten was missed for a different reason, so that the pattern of who got dug is itself a map of the genus's history. The small white lady's slipper of the prairie was effectively gone, plowed under with its grassland, before the trade ever reached its peak; the diggers never found it because it had already left. The Kentucky lady's slipper was not recognized as a distinct species until 1981, long after the trade was dead, so it was almost certainly dug all along in the southern woods and tossed in with the yellow root as "*pubescens*," collected by people who had no idea they held a different plant. And the western and far-northern species (the serpentine California lady's slipper, the boreal trio up past where any nineteenth-century collector network ran) were simply out of reach, on the far side of a continent the East Coast pharmacy never crossed. Same genus, twelve plants: the human record dug two of them nearly bare and never touched the rest, each spared by a different accident of geography or timing or taxonomy.

The yellow lady's slipper as a medicinal plant, from Millspaugh's American Medicinal Plants *(1887), in the era of the "American Valerian" root trade. Public domain.*

Poaching, the small sharp losses

The root trade is history. Poaching is now. It is a smaller pressure than habitat loss and a far harder one to document, because the typical case is not a courtroom but a hole in the ground where a plant used to be, found weeks after the digger has gone.

The Big South Fork theft I set down in the pink's chapter is the shape of it: about thirty pink lady's slippers gone from federal ground in a single June of 2020, and no one caught. The demand behind that hole is everyday. The pink moccasin flower is the lady's slipper people most want in their gardens, and because it is very nearly impossible to keep alive out of its acid ground, almost every one dug from the wild for a garden is simply a wild plant killed slowly in a flowerbed. And the pressure is not only American. The spotted lady's slipper, that circumboreal traveler, is globally secure across the top of the world, yet in South Korea it has been reduced to two known populations on a single mountain, driven there in large part by collectors digging it for its beauty. A hole in the ground in Tennessee and a hole on a Korean mountainside are the same small sharp loss.

Climate, honestly

Climate is real for these plants, and it is doing something subtler than simply cooking them, so it is worth getting right. The one genus-wide look we have is a study by Patrick Smallwood and Dorset Trapnell, published in 2022, that did something clever: rather than guess at the future, they compared where six of these species actually grew in the nineteen-eighties against where they grew in the twenty-tens, and asked what had already shifted. Two findings matter here. First, of all the variables they tested, the single most consistent predictor of where these orchids could live was the soil: a result that makes deep sense, because the soil is where the fungus lives, and the orchid can only go where its fungus already is. Second, the species did not all move the same way. The brownie's band of suitable climate shifted some five hundred kilometers (about three hundred miles) to the northeast, with the modeled area of good ground ballooning by a factor of twenty. The ram's-head's suitable climate slid the opposite direction, roughly three hundred and thirty-five kilometers (about two hundred and ten miles) to the west. The spotted lady's slipper's shifted south and uphill, an odd direction, since it was already pinned against the top of the continent with nowhere north to go.

Now hold those tidy numbers against the ground, because this is the whole danger. The brownie's modeled good ground expanded enormously; over the same span, surveys found something like forty-five percent of its real populations had blinked out, the losses concentrated at the warm low-elevation edge exactly as warming would predict. The ram's-head's biggest population on earth fell by about sixty-five percent. The map says expanding; the ground says dying. As we saw out west, the gap between the two is the threat: a lady's slipper disperses by dust seed that must land on exactly the right fungus to live, so a patch of newly suitable climate four hundred miles away is no refuge. Climate does not have to make the old ground uninhabitable. It only has to move the habitable ground faster than a fungus-dependent dust seed can follow.

There is a smaller, quieter climate signal too, in the timing of bloom. The pink lady's slipper in North Carolina, read across more than a century of pressed herbarium specimens from 1886 to 2022, has advanced its flowering by about three weeks. Its bumblebees keep their own schedules and are not all matching pace; several of the very bumblebees that pollinate it are among the most steeply declining bees on the continent. A flower that opens three weeks early into a world where its pollinator has not yet emerged is being slowly, invisibly uncoupled from the partner it cannot reproduce without.

And then the strangest threat of all, the one that runs backward from every other. The pink lady's slipper lives by sourness; it needs that pH-four ground. Across much of the eastern United States the soil is slowly growing less acid, recovering from the worst of twentieth-century acid rain as the air gets cleaner. For nearly every living thing that is good news. For this one orchid it may be a quiet erosion of the exact niche it took as its own, the sour ground sweetening underneath it as the sky heals. This is a reasoned worry, not a measured decline, and no one has caught the mechanism in the act. Here is a plant that could be undone not by poison in the air but by the *removal* of poison from the air. A condition withdrew. Even a healing one can pull a flower down.

And the rest, briefly

The other pressures are real but smaller, and they cluster around the same idea of a condition shifting. White-tailed deer, overabundant across much of the East, browse the queen hard enough that the long demographic studies in West Virginia by Marc Kéry and Katharine Gregg could measure it: across eleven years, a population grazed hard slid downward while one under moderate browsing held steady. Hydrology, the rise and fall of water tables, the draining and ditching of wetlands, sits underneath the fate of every fen and seep species. And invasive plants press in at the edges: at the Grand Sable dunes in Michigan, where the world's largest population of the ram's-head fell by roughly two-thirds, the main culprit was the woods quietly closing in as fire was suppressed, with invasive baby's-breath and off-road-vehicle traffic grinding at the colony from the side. None of these usually finishes a colony alone; each is one more clause under strain.

What keeping looks like

The small white lady's slipper in Illinois shows what the other half looks like. Listed as state-endangered in 1980, it was brought back through the patient, unglamorous work of finding populations no one had recorded, protecting the ground they stood on, and putting fire back onto the prairie with prescribed burns; it was downgraded to threatened in 1998 and taken off the list entirely in 2014. Up in Manitoba, the Tall Grass Prairie Preserve holds tens of thousands of stems of that same orchid, a stronghold built simply by keeping a piece of real prairie whole and letting it burn the way prairie burns. That is the hopeful half of the standing-agreement idea: if these plants vanish when a condition withdraws, they persist when the condition is kept, and a condition can be kept. How a nursery flasks a seed, how a recovery program brings a species back, that is the next chapter's work. The principle here is plain. Keep the agreement and the flower keeps itself: the sour ground sour, the fen wet, the fire on its old rounds, the canopy open enough to let the light down. It has been keeping itself, where the conditions held, for a very long time.

So go back, one more time, to the botanist on the hillside, reading the ground before she has seen the flower. The reading runs both directions now. It tells her where to look, and it tells her what the plant is owed — which rock, which water, which light, which fire. The morning I learned to half-do the trick myself, I was on a low ridge in old pine woods, the duff thick and slow and acid underfoot, the light coming down broken through the canopy, and I crouched at a spot that simply felt right before I knew why. The ground was sour. And there, where the slope caught the seep and the leaf litter lay undisturbed, was a single pink pouch on its bare stalk, nodding a little in the air, standing exactly where every condition said it should. I did not dig it. I read the rest of the hillside off that one flower, the way you read a sentence off a single sure word, and then I left it to keep its agreement, and walked on.

The Trouble with Wanting One

On a steel bench in a quiet lab there is a glass jar, stoppered, holding what looks like a smear of green jelly clinging to a slope of clear nutrient gel. Lean in close. The smear is not jelly. It is dozens, maybe hundreds, of tiny green bodies the size of pinheads, each one a lady's slipper at the very beginning of its life, fed not by a fungus and not by soil but by the sugar dissolved in the gel around it. No woods were disturbed to make this jar. No colony lost a single plant. Every green speck in it is a future garden orchid that was never dug, never carried home, never torn out of the ground where it was born. This is the hopeful end of a hard subject, and I want to start you here, with the jar, because almost everything else in this chapter is about the difference between the orchid in that glass and the orchid you might be tempted to lift out of the wild with your own hands.

Here is the difference. A wild lady's slipper is one of the very few things on earth you cannot have by taking it. Take it, and you do not get the plant. You get its death, slow and nearly certain, and a hole left behind in a wild colony. You can truly have one of these orchids only two ways: by leaving it exactly where it stands, or by *making* one, the way that jar was made. The wanting is not the problem. Wanting a lady's slipper is one of the most natural responses a person can have to seeing one. I have felt it myself. Years ago, on a low ridge in the oak woods of West Virginia, in the first warm week of June, I knelt in front of a pink lady's slipper standing alone in the leaf litter, the pouch flushed and faintly veined and almost obscenely alive against all that brown. I had a trowel in my pack. I remember the exact arithmetic of the wish: how easy it would be, how few minutes, how good it would look at home. My hand actually moved toward the pack. Then I made myself sit back on my heels and simply look at it. The longer I looked, the more it stopped being a thing I could own and started being a thing that belonged exactly there, fastened to that particular slope and that particular soil in ways I could not see and could not carry. The wanting did not vanish. It turned into something quieter, closer to gratitude, and I left the plant where it stood and walked on. The trouble with wanting one is only ever the trouble with *taking* one. Everything good in this chapter follows from getting the grammar of that desire right.

The trouble: why taking kills

Start with a number that should stop the impulse cold. Dig a wild lady's slipper out of the ground, carry it home, and plant it in your garden, and the odds it survives are poor. For the pink lady's slipper, the one most people dig, they are worst of all, somewhere around five percent. Roughly nineteen out of twenty die. Not because you did it wrong, though most people do, but because the thing cannot, in the ordinary sense, be moved at all.

To see why, you have to remember what a lady's slipper actually is, and I laid both halves of it out earlier in this book. The first half is the partner underground: these orchids do not feed themselves the way a tomato or an oak does, but live by a particular soil fungus threaded into their roots, and never fully wean off it, drawing a share of their living from it until the day they die. The second half is time: the better part of a decade, often much longer, from a dust-sized seed to a first flower, and the visible plant only the youngest, briefest part of an individual that may have been spreading and storing underground for decades, drawing down a rhizome reserve that, once spent, never refills.

Now try to dig that up. Your shovel cannot lift the fungus. The fungal partner is not a tidy clump of roots you can wrap in burlap; it is a network woven through the surrounding soil, specific to that spot, and it does not come along when you pull the plant. And your shovel certainly cannot lift the time. You cannot carry home the decades of accumulated underground growth that built the plant, any more than you can transplant the age of a tree by moving the tree. So what arrives in your flowerbed is not really a lady's slipper. It is a severed thing: a plant cut off from the partner it feeds through, running on the last of a reserve it can no longer rebuild, set down in soil whose chemistry and fungal life are wrong for it in a dozen quiet ways. It may green up the first spring on stored reserves and fool you into thinking it took. Then the account runs dry, and there is no fungus to refill it, and it is gone. A wild-dug lady's slipper in a garden is, almost always, a wild plant dying slowly, on a delay.

The cruelest example is the one people want most. The pink lady's slipper (*Cypripedium acaule*), the moccasin flower of half the woods in eastern North America, is far and away the lady's slipper that ordinary people recognize and covet, the one whose pink pouch says orchid to someone who has never thought about orchids in their life. It is also very nearly the most impossible of all of them to keep alive once moved. It is bolted to extremely acid ground, the sour pine-woods soil I described at the pink's own chapter, soil down around a pH most garden plants would refuse, and it is exactly there, in that narrow chemistry, that its fungal partnership holds. Lift it out of that, set it in ordinary sweet garden loam, and you have broken the one condition the whole arrangement was built on. A wild-dug pink lady's slipper in a flowerbed typically dies within about three growing seasons. People dig them every year anyway, in good faith, charmed and meaning no harm, and they kill them every year, one transplant at a time.

And the killing does not stop at the one plant in your hand. Every wild lady's slipper taken is subtracted twice. There is the slow death in the garden, which you see. And there is the hole left in the wild colony, which you do not, because you have already driven away. The Big South Fork theft I described earlier is the shape of it: thirty pink lady's slippers gone from public ground in a single June, leaving holes where the plants had been. Thirty plants out of a wild population is a wound. Thirty plants into gardens is, within a few years, thirty dead orchids and nothing to show. The taking fails twice over: it does not give you the plant, and it does take it from the woods. The wanting was human and fine. The taking was the whole trouble.

The law: a fence with one gate

There is a law around all of this, and it is worth knowing honestly, neither more frightening nor more reassuring than it actually is.

The headline fact is older than most readers expect. Every lady's slipper on earth, the whole genus *Cypripedium*, has been listed on Appendix II of the international wildlife-trade convention, the agreement most people know by the shorthand CITES, since the first of July, 1975. So has the entire orchid family, in fact. Appendix II is the convention's middle tier; what it mainly governs is international trade, requiring permits to carry wild-collected plants across borders, rather than reaching down to regulate the plant in your local woods. It governs borders, not woodlots. It is a fence built around the global market in these orchids, and it has stood for half a century.

It helps to see where our native slippers sit on that fence relative to their famous tropical cousins. Back at the start of this book I closed the door on those cousins: the showy slipper orchids of tropical Asia and the American tropics, *Paphiopedilum* and *Phragmipedium*, the ones whose hybrids fill the benches at orchid shows and whose pots at the garden center say "lady slipper" on the tag though not one of them is native to our woods. Those two genera sit on the stricter tier, Appendix I, and have since 1990. That difference matters in a practical, almost comic way: the "lady slipper" you would actually buy at a garden center is almost certainly a tropical *Paphiopedilum*, governed more tightly than any of our wild natives, and it is a houseplant, not one of ours.

The single most important thing CITES does for the subject of this chapter is something it explicitly chooses not to regulate. A sealed sterile flask of orchid seedlings, the kind of jar I opened this chapter with, grown in a closed laboratory container, is exempt from the convention. The framers reasoned, sensibly, that controlling trade in flasked seedlings from closed nursery systems has little to do with protecting wild populations, and that burdening it with paperwork would only discourage exactly the kind of propagation that takes pressure off the wild. That exemption is the open gate in the fence, the one deliberately left for the orchids that were made rather than taken, and it is why a sterile flask of seedlings can move from a lab in Germany to a grower in Minnesota to a gardener's bench without a wild plant ever entering the story.

Now the part that surprises people, because it cuts against the assumption that anything this precious must be wrapped in protective law. Of the twelve native lady's slippers of North America, not a single one is listed under the United States Endangered Species Act. Not as endangered, not as threatened, not even as a candidate. The California lady's slipper (*Cypripedium californicum*), pinned to its serpentine seeps and rated Endangered on the international red list, carries no federal protection in its own country beyond that genus-wide CITES listing. Its federal status page reads, like all the others, "Not Listed." Only one native lady's slipper anywhere has a national endangered-species listing in either country that shares this flora, and it is across the border: the small white lady's slipper (*C. candidum*), the candle-white prairie orchid, listed under Canada's Species at Risk Act, where it was first placed in 2003 and reassessed to threatened in 2014. Eleven of the twelve, in other words, lean on a half-century-old trade treaty and a patchwork of state and provincial collection laws, and on very little else.

There is a stronger enforcement tool in the United States, and the story it tells is quietly the most telling fact in this whole legal picture. The Lacey Act, the country's principal law against trafficking in wildlife and, since a series of amendments, in plants, carries real teeth: knowingly trafficking a protected plant for sale can be a felony, with serious fines and the genuine possibility of federal prison. People have gone to prison for slipper-orchid trafficking under it. But here is the thing. Every one of those prosecutions has been about the tropical slippers, the *Paphiopedilum* and *Phragmipedium* smuggled out of Asia and South America. The best-known cautionary tale in the whole specialist community is a Texas hobbyist named George Norris, who ran a small import business in tropical slippers and ended up serving seventeen months in federal prison in the early two-thousands for paperwork and permit violations. Search the record for a single case in which a native North American *Cypripedium,* one of our own twelve, was the charged plant, and you will not find one. The body of slipper-orchid case law exists. It has simply never once been written about a native species. Make of that what you will: it may mean our natives are protected by their difficulty more than by any statute, or it may mean the small steady losses to the trowel are exactly the kind that never reach a courtroom. Both are probably true.

And one current event, for honesty's sake, because it is the kind of thing that sounds like it should have mattered and did not. The convention's member nations met in late 2025, in Samarkand, for the regular conference where species get moved between the appendices. Going in, the United States had floated, on an internal pre-decisional list, the idea of uplisting three western and boreal lady's slippers to the stricter Appendix I. By the time the meeting's official proposals were filed, those three had been quietly dropped, never formally submitted. The conference came and went. For our twelve lady's slippers, nothing changed. The fence is exactly where it was in 1975, with the same one gate left open for the made plant.

The right way to have one: making, not taking

So you want a lady's slipper, and now you know you cannot have a wild one in any meaningful sense. Good news: you can have a made one, and the making is itself one of the most quietly astonishing tricks in horticulture. It is worth understanding, because understanding it is what lets you buy one well.

Recall the orchid's deepest problem, the one the whole first half of this book circled. A lady's slipper seed is dust, packing essentially no food, and on its own it cannot germinate; it must be found in the dark by the right fungus, which threads in and feeds it the sugars it cannot make for itself, for years, before a single leaf appears. That is the bottleneck. That is why these plants are so rare, so slow, so impossible to casually grow. And here is the trick that gets around it. You give the seed the sugar yourself. You sow it not in soil but on a sterile gel, a chemically defined medium worked out to supply the carbon and minerals a fungus would otherwise hand over, in a closed sterile container where no competing mold can get in. The seed, fed, germinates without ever meeting a fungus at all. The method has a name: **asymbiotic** germination, without the symbiosis, growth deliberately uncoupled from the partner. In the simplest terms, the petri dish replaces the fungus.

The lineage of that idea reaches back further than you might guess, and it touches this book in a way I find lovely. The basic breakthrough, that orchid seed could be germinated on a defined nutrient medium with no fungus present, belongs to an American researcher named Lewis Knudson, who published it in 1922. The refinement that made it work for the notoriously difficult lady's slippers specifically came two decades later, in 1943, from a botanist named Curtis. That year, 1943, and that name should ring a faint bell, because it is the same Curtis whose patience opened the chapter on the long wait, the one who sat with these orchids long enough to document just how many years they take to climb from seed to flower. The same careful attention that measured the orchid's patience also learned how to feed its seed by hand. The man who clocked how long a lady's slipper makes you wait is the man who taught the petri dish to do the fungus's job.

There are a couple of refinements worth knowing because they show up in how plants are sold. The first growers call the **green-pod** technique: instead of waiting for a seedpod to ripen and harden, you harvest it while it is still green and immature, a few weeks after pollination, and that not-yet-hardened seed germinates far more readily than fully mature seed does. It is the basis of most modern lady's slipper propagation. The second is that the hardest species, the ones that simply refuse to grow on the basic recipe, can sometimes be coaxed along by adding a plant hormone to the gel, a chemical nudge toward growth. These are the kitchen secrets of a strange and patient craft.

None of it is fast. From sowing a seed to a plant big enough to flower in a garden takes, in round numbers, six to ten years, the laboratory shaving only the front of that off the wild timeline. And there is one stretch of those years where nearly all the dying happens, the single hardest moment in the whole process: the hand-off from the sterile flask to actual garden soil, the moment growers call **deflasking**. Inside the jar the seedling has been protected, fed, kept free of every competitor. Then it is moved into ordinary dirt, where it must find the partner it never had or do without it, exposed to the molds and the weather of the real world. Most of the loss in the entire enterprise is concentrated right there, at the door between the glass and the ground. That difficulty is exactly why a properly grown lady's slipper is not cheap, and it is the hinge of the whole buyer's ethic.

Which brings us to the practical heart of this, framed not as a lecture but as a single useful skill: how to want one responsibly. There is a rule of thumb in the trade, and it is the most useful sentence a would-be lady's slipper buyer can carry. It comes from Tony Avent, a well-known North Carolina nurseryman, and it goes like this: "When you see flowering-size plants sold for less than $35, you can be pretty sure that they have been wild collected." A lab-grown lady's slipper represents the better part of a decade of skilled labor and a brutal survival bottleneck; it cannot be produced cheaply, and no honest grower can sell a blooming one for pocket change. So a flowering-size plant offered for less than about thirty-five dollars is, with rare exceptions, a plant that cost the woods instead. Cheap means dug. That single price floor is the most reliable thing a beginner can use, and it leads straight to the one real rule: buy only stock that was raised from flask or from seed, propagated rather than collected, and ask the seller plainly which it is. An honest grower will be glad you asked.

I should be honest about how short the available list is, because it is part of the picture and it is humbling. Of the twelve native North American lady's slippers, only four are realistically available, in 2026, as ethically propagated plants a gardener can actually buy: the showy lady's slipper (*C. reginae*), the queen, which has become the genus's great propagation success story; the yellow lady's slipper (*C. parviflorum*), the easiest of the natives and the workhorse of the labs; the big Kentucky lady's slipper (*C. kentuckiense*); and the spotted lady's slipper (*C. guttatum*). Four out of twelve. The pink, the very one most people want, is the wall: the single hardest of all of them to propagate, the one whose flask-grown seedlings most reliably fail to make the jump to soil, and the contemporary effort to crack it is real but not yet finished. And the small brownie lady's slipper (*C. fasciculatum*), the plain dark one of the western mountains, sits at the far limit: there is no reliable method for it at all, no one is producing it, and you genuinely cannot buy one, propagated or otherwise, at any price. The demand for the eight you cannot buy honestly is exactly the demand that drives the trowel into the wild. For those eight, the wanting has nowhere honest to go yet, and the plant stays where it is.

For the four you can buy, though, the wanting has an honest place to go. On the porch and in a shaded bed by the house grow a showy lady's slipper, nursery-propagated, bought as a small plant, and a garden hybrid with the big Kentucky lady's slipper somewhere in its parentage, crossed long ago with an Asian cousin and grown on by the thousand in nurseries that never touch a wild plant. Neither came out of any woods. I did not dig either one, and I never will.

The author's own Cypripedium *'Philipp Dark,' a nursery-bred hybrid of the Kentucky lady's slipper, blooming on the porch. Bought as a propagated plant, never dug from the wild. Photograph by the author.*

The people doing the making are worth naming, lightly, not as a shopping list but because the work has faces. In northern Minnesota, a small operation called Spangle Creek Labs, run by Bill and Carol Steele since 1990, grows native lady's slippers from seed by the asymbiotic method and sells seedlings to gardeners patient enough to raise them. In Massachusetts, Hillside Nursery, where Amy Murray has carried on the lady-slipper work since 2005, offers flask-grown plants and hybrids. In Bavaria, a nursery carrying the name Frosch, after the German breeder Werner Frosch who helped pioneer hardy lady's slipper hybrids and co-wrote the standard book on them, has been run since 1994 by Michael Weinert, producing the hybrid lady's slippers that thrive in cold gardens. And at the Smithsonian's research center on the Chesapeake, a program called the North American Orchid Conservation Center works on the harder, deeper problem of propagating natives that no business can yet grow at scale. These are the makers. Between them they are slowly turning a desire that used to mean a dug wild plant into one that can mean a flask-grown one instead. That is the substitution that matters: demand met by making rather than taking.

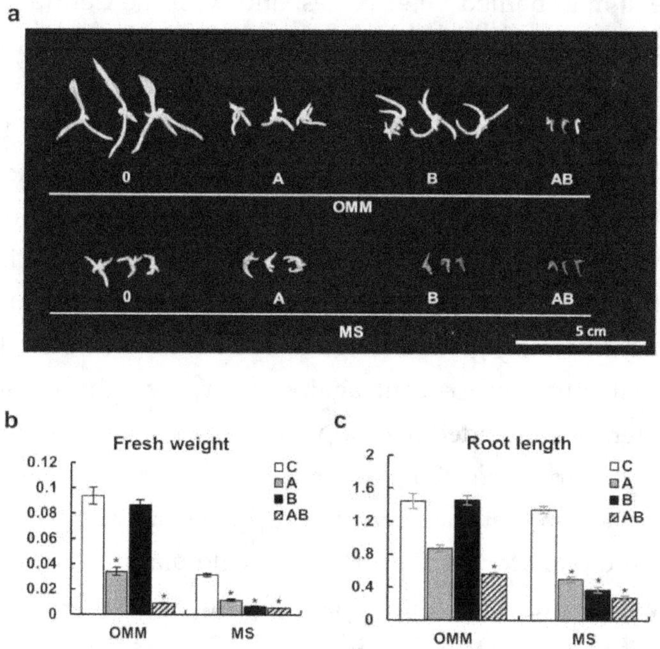

Lady's slipper seedlings raised in a sterile flask on nutrient gel: the patient lab method that lets you keep one of these orchids without taking it from the wild. (Shown here in Cypripedium guttatum.*)*

The hopeful human heart

I want to end on the best version of all of this, because it is also true, and because it answers the wanting with something better than a warning.

In 2004, a high-school student in Shreveport, Louisiana, named Kevin Allen was out in the Kisatchie National Forest, the only national forest in the state, six hundred thousand acres of piney woods, when he found something almost no one alive had seen there: a Kentucky lady's slipper, the big ivory southern one, and it was carrying seedpods. That mattered enormously, because on all six hundred thousand acres of that forest, the Kentucky lady's slipper was down to two locations and a combined total of five plants. Five. A species hanging by a thread you could count on one hand. And here was a teenager who had found one of them setting seed.

What he did next is the whole point. He did not dig it. The next year, in 2005, he partnered with the Forest Service botanist on the Kisatchie, a man named Peter Nilles, and with the Central Louisiana Orchid Society, the local hobbyists. The precious seedpod went off to Spangle Creek Labs in Minnesota, to be germinated by the asymbiotic method, the petri dish standing in for the fungus that the Louisiana woods could no longer reliably supply. Grant money paid to grow the seedlings on. And the survival numbers climbed, batch over batch, as the growers learned the plant: from twenty-five percent in the first attempt to seventy-one percent in a later one. And in a separate trial, the seedlings deliberately given a fungal partner survived better than those left without, fifty-two percent against thirty-five, the made orchids doing better when handed back a piece of the relationship the lab had let them skip. From five wild plants and one student's restraint, a future for the species on that forest, grown one flask at a time. Nobody had to lose a wild plant for it to happen. One wild plant, left where it stood and allowed to finish making its seed, became the source of all the rest.

It is not the only story like it. In New Hampshire, middle-school students at a small school called Crossroads Academy, working with the New Hampshire Academy of Science and the state's orchid society, have learned to flask lady's slipper seed themselves, and have grown hundreds of tiny showy lady's slipper protocorms, those same green pinheads from the opening jar, raising them from dust on sterile gel. They have taken their results to a World Orchid Conference and held their own among the professionals. Children, doing by hand the trick that Knudson and Curtis worked out, feeding orchid seed the sugar a fungus would have given it, bent over the flasks in a school lab where the rarest flower in the woods is starting from dust.

And underneath all of it sits one last fact, the deepest in the chapter, and it turns the whole subject back toward wonder. Ask why we can make only four of the twelve, why the pink is a wall and the brownie a flat impossibility, and the answer is not really about gel recipes or hormones or patience. It is about the fungi. A researcher named Melissa McCormick, at that Smithsonian center, has reported a staggering thing from the work of culturing the fungal partners of native orchids: of the orchid fungi sequenced so far, something on the order of ninety-nine point nine percent appear to be species new to science. New to science. We are trying to grow plants whose essential partners we have, for the most part, not yet even named. The reason we cannot yet make most of these orchids on demand is, at bottom, that we still do not know the fungi that feed them. The propagation gap is a knowledge gap. We cannot reliably make what we cannot yet feed, and we cannot yet feed what we have not yet met.

So the jar on the bench is not the end of the mystery. It is a few species' worth of solved problem floating in a sea of unsolved one, a handful of orchids we have learned to make set against eight we have not, against a hidden world of fungi we have barely begun to read. The right way to have a lady's slipper turns out to be the same as the right way to regard the whole genus: with patience, and with the humility of someone who knows how much is still unknown. Kevin Allen, seventeen years old in the Louisiana woods, got it exactly right with the first move. He found the rarest thing on the forest, and he left it rooted where it grew, and he asked it, in effect, for its seed instead of its life. The orchid is still there. So, now, are its children, coming up green in the dark.

How to Find Them Without Loving Them to Death

A hiker stops on a trail in late spring, somewhere in the eastern woods, because a pink lady's slipper has caught her eye low among the leaf litter, a single veined pouch nodding on its bare stalk. She reaches for her phone. She does what almost anyone would do now, holds it up, and frames the flower on the screen. The phone is a camera, so it makes a photograph. The phone is also a small computer running an app that knows orchids, so in about two seconds a name floats up under the image: *pink lady's slipper*, with a little green check of confidence. And the phone is a third thing too, a satellite receiver that always knows where it is standing, so quietly, automatically, without her asking, the app has stamped onto that photograph the exact coordinates of the plant, accurate to a few steps, and stands ready to share the whole package with the entire internet at the tap of a button.

Think about everything that just happened in those two seconds. She found the orchid, identified it correctly, made a permanent record of it, and located it on the surface of the earth to within a few meters, using one device she already had in her pocket, with no skill required beyond pointing. None at all. Two seconds, no skill, and the flower is named and mapped. I do not want to wave that away. It is also, for this particular flower, the single most dangerous moment in its modern life. The same pin that lets a scientist count it lets a thief drive straight to it. Everything in this chapter lives inside that contradiction, and learning to hold both halves of it at once is most of what it means to love one of these orchids without loving it to death.

This is a chapter about what to do once you have found one. I spent an earlier chapter on the skill of finding them at all, the way a good field botanist reads the rock and the water and the light and predicts the flower before either party sees it, and I will not redo that here. Assume you have found your lady's slipper. The orchid is in front of you, real and improbable and yours to look at. The question this chapter answers is simply: now what? What do you do with your two feet, your camera, and the record you are about to make, so that the finding is a gift to the plant and not the beginning of its end?

The most dangerous thing about a rare orchid is its address

Start with the thief, because understanding the thief is what makes the rest of this make sense, and because the thief is not who you picture.

You might imagine that stealing a wild orchid takes some dark expertise, the way safecracking does, some hard-won criminal skill that most people simply do not have and so most plants are safe. It takes nothing of the kind. The one genuine obstacle standing between any given plant and the trowel is not skill or nerve or equipment. It is knowledge. The poacher needs exactly one thing, and only one thing: he needs to know where the plant is. Everything else about what he does is furtive and small and best done where no one is looking, which is precisely why the secret of the place is the plant's true defense. A rare orchid is kept safe not by being hard to dig but by being hard to find.

Now look again at what the phone did on the trail. When the app stamped those coordinates onto the photograph, it did not record a vague region or a pretty hillside. It recorded an address, precise enough to walk to, the one and only thing a would-be thief lacks, and the one thing he cannot easily get any other way. A geotag on a rare-orchid photo is not metadata. It is a set of directions to a living plant that cannot move, posted where anyone can read them. Earlier in this book, out in the western mountains, "address" meant the place a plant is welded to and can never leave, bolted to one slope and one soil and one fungus. Here the address is the thing on the map, and it is the very fixity that pins the plant to its ground that makes a leaked coordinate fatal. A deer can run. A bird can fly. A lady's slipper can only stand exactly where its coordinates say it stands and wait for whoever comes.

This is why the tools that make all this possible have been quietly built to protect against themselves, and the fact that they have is one of the more hopeful things I know about the modern naturalist's world. The largest of these platforms, iNaturalist, the one whose app named the flower on the trail, does something deliberate the moment a species is flagged as sensitive. When you log a rare lady's slipper, the platform automatically smears its location. It does not publish your precise pin. It shows the observation sitting somewhere inside a wide blurred circle, often several kilometers (a few miles) across, sometimes more, so that the record exists and counts and feeds the science while the actual plant stays hidden inside the haze. The setting that does this is called **geoprivacy**, and for the rarest of these orchids a curator can switch it on for the entire species at once, so that every observation anyone ever logs is blurred by default. The pink lady's slipper (*Cypripedium acaule*), the most photographed of all our native slippers by a wide margin, fills the maps with its records; the showy lady's slipper, rarer and more coveted, is among the ones most commonly hidden behind that automatic blur. The platform built the fog into itself on purpose.

The state natural-heritage programs, the official keepers of where rare plants grow, have done the same thing for far longer, the careful **obscuring** of sensitive locations being one of the oldest disciplines in the trade. Ask a heritage program for the location of a rare orchid and you will get a county, a watershed, a generalized dot, never a pin you could walk to, unless you are a vetted researcher with a reason. They learned, decades ago, that a precise location is the most dangerous single fact they hold about a rare plant, and they treat it accordingly.

For a rare orchid, its location is the most dangerous thing about it. Not its rarity, not its beauty, not its market price. Its address. The flower itself you can photograph, admire, post, and celebrate to your heart's content; none of that hurts it. The one piece of information that can get it killed is the precise spot where it stands. So the whole ethic of finding these plants reduces, in the end, to a single clean discipline: you may keep everything about the encounter except the coordinates. You take the picture and leave the pin. You can love the orchid out loud. You just cannot tell the internet exactly where it lives, because the internet is not the right keeper of a rare plant's address. The state heritage program is.

And the deeper question of who gets to govern knowledge about a place has a far older answer than anything the heritage programs invented. The discipline of confidentiality I have described, the careful keeping of a rare plant's address, runs parallel to Indigenous frameworks for data and place that long predate it: the CARE Principles for Indigenous Data Governance, and the work of organizations like Local Contexts, which hold a community's right to govern knowledge about its own lands and living relations. There is a plain corollary the whole field is still learning to honor, that a lady's slipper population on tribal land is named, located, and discussed only with the explicit permission of that Nation, on that Nation's terms, never by default. I am a settler writing from outside these frameworks, pointing at them rather than speaking for them, and I name them here only to say that the instinct this chapter is built on, that some locations are not yours to broadcast, is one that Indigenous peoples articulated as principle long before it became a checkbox in an app.

How a sensitive record is protected. The true location is buried inside a wide blurred circle, so the observation can exist without handing anyone the plant's address.

At the plant: the danger is underfoot

Before the record, though, there is the encounter itself, the few minutes you spend standing in front of the flower, and there is a real way to harm a lady's slipper in those minutes that has nothing to do with the trowel and nothing to do with coordinates. It is gentler than poaching and more common than poaching, and most people who do it never know they did anything at all. The danger at the plant is not in your hands. It is under your feet.

Remember what is actually down there, because I built most of this book around it. Look down. The visible flower is the smallest, briefest part of the organism. Just under the surface runs the rhizome, the thick horizontal stem that is the plant's whole savings and memory, sitting shallow, a finger's depth or two below the leaf litter. Spreading out from it are the roots, and threaded through the soil all around those roots, out past the plant in every direction, is the fungus, the partner this orchid feeds through and cannot live a season without. The living lady's slipper is not the pouch on the stalk. It is a shallow, fragile, invisible web of rhizome and root and fungus spread through the top few inches of ground in a circle wider than the plant, and you are standing on the edge of it, or in the middle of it, the moment you walk up to admire the flower.

So consider what happens when you kneel for a better photograph, or step in close to fill the frame, or shift around to the side where the light is better, or bring three friends over to look. The flower is untouched. But the ground around it is being compressed under your weight, the air squeezed out of the soil, the shallow rhizome pressed on, the fungal threads in the duff crushed and broken where your knees and boots come down. You can do real and lasting harm to a lady's slipper, the kind it may not recover from, without ever touching the part you came to see — and the damage is worse for being invisible and additive, each visitor certain they were careful, the compression adding up season after season until one year the plant simply does not come back.

Which turns the etiquette at the plant into something simple, and not at all a long list of rules. Keep your distance. Give the plant a wide berth, a generous circle of untrodden ground, and make your photograph from outside it, from where you are already standing on the trail rather than from a spot you had to crush your way into. Stay on the established path; the path is where the soil is already compacted and already dead to the orchid, which is exactly why it is the safe place to stand. And resist every small temptation to improve the shot. Do not pull away the surrounding leaves or snap off the twig that crosses the frame or clear the little screen of grass in front of the pouch; that "grooming" strips the plant of cover and tramples its root zone for the sake of a cleaner picture. Do not bend the flower toward you or turn it for the light. And it should go without saying by this point in the book, though I will say it, do not pick it, and do not dig it. The best photograph of a lady's slipper is the one taken from where your feet already are, of a flower you have not rearranged, with the ground around it exactly as undisturbed as you found it. A picture costs the plant nothing. It is the getting closer that costs.

Photographing from a respectful distance, on the trail, the leaf litter between camera and flower left undisturbed. The plant's roots and fungus live in that ground.

Looking as a contribution: a hundred years of it

Now the record, and the part of this that turns a private moment of wonder into something genuinely useful, because here is the small electric fact underneath modern orchid-watching: your photograph, logged the right way, is data. A flower you noticed on a Sunday walk and recorded with a blurred location becomes one more dot in a continental dataset about where these plants are, when they bloom, and how both are changing. Looking, done with a little discipline, is a contribution to the science. You do not need a degree or a permit. You need attention and a willingness to record what you see.

This idea feels new, born of the smartphone and the app, and in its mechanics it is. In its spirit it is about a hundred years old, and the taproot is worth knowing, because it tells you that the practice you are being invited into is not a tech-company novelty but the latest turn of a long American tradition of paying attention. It runs through a woman named Anna Botsford Comstock, a leader of the nature-study movement and the first woman to hold a professorship at Cornell, who in 1911 published a thick green book called the *Handbook of Nature Study* that became the bible of that movement. Comstock and the nature-study movement argued something that sounds obvious now and was not then: that ordinary people, schoolchildren most of all, should go outside and look closely at living things, not to extract anything, not even necessarily to become scientists, but because attentive looking was a worthy end in itself.

The line from that green book runs straight to the app on the trail. When you stop, look hard at a lady's slipper, and add your careful observation to a shared record, you are doing in 2026, with a satellite in your pocket, very nearly the thing Comstock was asking a schoolchild with a notebook to do in 1911: look closely, write down what is true, and understand that the looking itself is a kind of contribution. The tools changed beyond recognition. The proposition did not.

There is one footnote to the generous act that most people miss, and I include it because it is exactly the kind of small true thing that makes the difference between a record that helps and a record that helps more. When you upload a photograph to iNaturalist, it attaches a license to your image, and for a great many users the default is CC-BY-NC, Attribution-NonCommercial: anyone may reuse the photo, but not for commercial purposes. That sounds generous, and it is, but it has a quiet cost. A photo under that license cannot be used on the free encyclopedias, and it cannot be used in a book like this one. If you would like your pictures to be fully useful, reusable by anyone for anything, including the projects that most need good images of rare plants, you switch once to one of the more open licenses, CC0 or CC-BY, and from then on your uploads carry it; you can even relicense your whole back catalog in a single pass. One toggle does it, a thirty-second change in a settings menu that most contributors never open, and it turns a good photograph into a freely useful one.

Where to send what you find

If looking is the contribution, the next question is practical: where does the looking go? There is a real and slightly bewildering ecosystem of programs that want what an attentive amateur can supply, and the honest map of it is worth a few paragraphs, kept light, with the addresses and the sign-up details pushed to the back of the book where they belong.

The first and best fact is that of all the things an amateur can contribute, the single most valuable is also the most patient, and it is the one I would point you toward before any other. It is **phenology**: the timing of a plant's year, when it buds, when it opens, when it sets seed. The way you contribute it is almost absurdly simple. You find one lady's slipper, and you go back to it. Week after week across a season, the same plant, and each time you note where it is in its year. That is it. That returning, that willingness to visit one flower again and again rather than chase many, produces exactly the kind of data that is thinnest for these orchids and most needed — because the great signal everyone is now trying to read in the wildflowers is the slow advance of bloom time as the climate warms, and you cannot read a shift in timing without people patiently logging the timing, year over year, plant by plant. I have written elsewhere in this book about what that warming is doing to these flowers and the bees that depend on them; I will only say here that the studies that caught it were built out of exactly this kind of humble repeated looking, and that more of it is wanted. The amateur who adopts one plant and faithfully watches it through its season is doing the most useful thing on this entire list.

Then there is the deep mystery you can actually help solve, and this is the one that should make any patient watcher's pulse pick up. The Smithsonian's North American Orchid Conservation Center runs a participatory effort called the Native Orchid Pollinator Survey, led by a scientist named Melissa McCormick, that puts motion-triggered cameras at orchids to record who comes to pollinate them. Why does this need doing? Because of a fact that still astonishes me every time I say it: for roughly half of all North American orchid species, no one has ever documented the pollinator at all. Half. We do not know, for a great many of these flowers, what insect carries their pollen. And the lady's slippers with the emptiest records, the ones whose visitors have never been watched, are precisely the priorities. Which means an ordinary person with patience and a camera, sitting quietly at the right orchid, can contribute a piece of genuine discovery, can be the first human being ever to document who pollinates a particular flower. That is not make-work dressed up as participation. That is a real frontier with room on it for amateurs.

For the orchid you have actually found, the rare one standing in front of you, the right destination for its location is not a public feed but your state natural-heritage program, the official keeper I mentioned earlier, the one that knows how to hold a sensitive address safely. Every state with rare lady's slippers has one, and there is a single universal place to start finding yours, a public reference called NatureServe Explorer that will tell you what is rare where you are and point you to the program that tracks it. Report your rare-orchid sighting there, to the keeper, and it goes into the real conservation record without ever becoming a set of public directions.

A few of these volunteer programs are worth naming, lightly, because the work has faces and because seeing how they are built tells you what serious amateur monitoring looks like. In Wisconsin, a volunteer Rare Plant Monitoring Program trains ordinary people to track the state's listed plants, on the order of three hundred species of them, the ram's-head and the small white lady's slippers among the orchids on its list. In New England, the Native Plant Trust runs what is, as far as I know, the oldest rare-plant volunteer-monitoring program in the United States, its Plant Conservation Volunteers walking transects and counting rare plants since 1993, the work long shaped by the botanist Arthur Haines. These are not casual operations. They are the structured, trained, decades-deep version of the same impulse the schoolchild had with the notebook, and they are always looking for patient people. The full roster, with everything you would need to join one, is in the back of this book.

When the news is bad

Sometimes you will not find a flower. You will find a hole, or a patch of churned ground, or a row of cut stems, the signature of someone who came before you with a trowel and no scruples. It is worth knowing, calmly and in advance, what to do, so that the discovery does not just sit in your chest as helpless anger.

There is a national tip line for wildlife crime, run by the United States Fish and Wildlife Service, and the number is easy to remember: 1-844-FWS-TIPS. You can call it about a dug orchid, and you can stay anonymous if you want to. Often, though, the faster call is to a state conservation officer, the local game warden, simply because that person is closer to the ground and can get to the site sooner. Neither call is dramatic. You are just reporting what you saw, the way you would report any other theft, and letting the people whose job it is take it from there.

A remote orchid population, out in some national forest or roadless fen, may get formally counted by an actual botanist once a year, if that. The rest of the year, the only eyes on it belong to whoever happens to walk past. Which means a regular visitor, an ordinary person who returns to the same patch and would notice if something were wrong, is very often that population's only real monitoring. The starkest proof of this is a theft I told earlier in the book and will not retell, except for the one detail that belongs here: the reason anyone knew those plants had been dug, the reason the loss was caught at all rather than vanishing unrecorded, was that someone had counted them two weeks before. The count was the only witness. Without it the hole would simply have been the way things were. Your repeated, attentive, undramatic presence at a place is, more often than the professionals would like to admit, the entire monitoring program.

The plant that lives under guard

I want to end with a cautionary picture and then a small ordinary gesture, because the picture shows you the far end of what happens when a location gets out, and the gesture shows you how to make sure it does not.

The picture is in England. The lady's slipper there is not one of our twelve; it is the European yellow lady's slipper (*Cypripedium calceolus*), the sister species I have used all through this book as the old-world reference, the one Darwin set aside. For most of the twentieth century Britain believed it had lost the plant entirely, dug and picked out of existence by Victorian collectors and gardeners. Then, in 1930, a single wild plant was found, surviving alone on a hillside, the last of its kind in the whole country. The people who found it understood immediately what its location was worth, which is to say how dangerous that location was, and they did the only thing that could keep it alive: they told almost no one. Its whereabouts were held as a close secret for something like forty years. And in the end, even that was not quite enough. The secret eroded, as secrets do, and the threats came, and so today that last wild English lady's slipper spends its life under guard, watched around the clock through its blooming season, with periodic police protection, the rarest flower in Britain, caged against the deer and slugs as well. A wild orchid that has to be watched and walled, because its address could not be kept quiet enough. The plant whose location got out is the plant that now has to be defended every hour it is in bloom.

A wild Cypripedium calceolus, *the European cousin of our natives. In England a single plant grew so rare it had to be watched and guarded through every bloom.*

That is the worst case, and our twelve are nowhere near it, and I tell it not to frighten you but to show you with perfect clarity what the stakes of a location are. A round-the-clock guard is what the failure to keep an address looks like, taken to its limit.

So here is the gesture, the small undramatic thing that is the opposite of the guard, and it is where I will leave you. You are back on the trail, in front of your own lady's slipper, the real one, on whatever spring afternoon you find it. You have stayed on the path. You have given the plant its wide circle of untrodden ground. You hold up the phone and make the photograph, the flower exactly as you found it, unbent, ungroomed, the litter around it undisturbed. The name floats up; the little pin drops with it, those few meters of precise and dangerous truth. And then, before you do anything else, you find the setting and you tap the toggle, and you watch the location go soft, the precise point dissolving into a wide blurred circle that could be anywhere across a few kilometers of woods. The record is saved. The orchid is counted. Its address is yours alone, and the plant's. You put the phone back in your pocket, and you take one last look at the flower, and you walk on, and you leave it where it is.

The Long Game

There is a pink lady's slipper (*Cypripedium acaule*) I have been visiting for most of my life. It grows on a low rise in a stand of white pine and oak near home, on ground sour enough that not much else crowds it, and the first time I knelt in front of it I was young enough that the kneeling cost me nothing. I keep going back. Nearly every spring, in the last warm days of May, with no tag staked at its base and no notebook, I climb the rise and look for the bare scape between the two pleated leaves, and find it, more years than not.

I want to be careful about what I am and am not claiming, because a plant like this invites a person to make it into a story about himself, and that is exactly the wrong way around. I cannot prove it is the same plant. A pink lady's slipper spreads underground and sends up shoots that come and go, and what I greet each spring may be the original individual or a shoot of it or, some years, no flower at all, only the leaves, or nothing, the plant down in the dark riding out a season it decided to skip. I have learned not to count on finding it. Some years I have stood on that rise and seen only litter and moss where I was sure it would be, and walked home convinced I had finally lost it, and found it again the next May exactly where it always was. It does not keep my schedule. It keeps its own.

What I can tell you is this. That flower was opening on that rise before I knew it existed, and it has gone on opening through every change the years have made in me, indifferent to all of it, and the arithmetic of that has slowly rearranged how I think about nearly everything. I have aged. The plant, in any sense that matters, has not. I have been in a hurry for most of my life. The orchid has never once been in a hurry. I came to it, at the start, wanting to know what it was. I keep going back, now, for something harder to name, which is the thing this whole book has been circling and which I want to try to gather up here at the end. Call it the long game. The orchid has been playing it the entire time, and I have only lately understood that I was being taught.

What the flower was running

Let me gather the threads, because by now you are holding all of them, and a close is not the place to learn anything new. It is the place to feel the whole shape of what you already know.

Go back to the very first thing I showed you, kneeling in the leaf litter: the pink pouch that looked, to the untrained eye, like a flower trying to be beautiful, and turned out to be a flower running a con. You know how it works now. The color and the faint sweetness are an advertisement for a meal that is not there. A bee is drawn in, finds no nectar, and is let out only by the single forced route the flower has built for her, past the stigma and then the anther, so that she pays for her disappointment by carrying the orchid's pollen to the next plant she is fooled by. The beauty was never decoration. It was a tool, tuned over deep time to the eye of a particular bee, and the fact that it stops us in our tracks too is a coincidence the plant knows nothing about.

And you know the rest of the trap's cleverness, which I spent a whole chapter on and will only touch here. The pouch does not merely catch a bee. It sizes her. The inrolled slit down the front of this pink one admits the strong and turns the weak away; the round-mouthed pouches of its cousins let anyone in and let only the right body out the working door. The trap is really a sieve, choosing rather than catching, and the choosing is the point. Visited by many, pollinated by few, and the fewness is not a failure in the machine. It is the machine.

I am not going to re-explain any of that. I bring it back only so you can hold the whole con in your hand for a moment and see what kind of thing it is. It is patient. It gives nothing and waits, gorgeously, for the rare naive visitor, season after season, banking on the long odds because the few good visits are worth more than a thousand cheap honest ones. The con is the work of a lineage that has had something like thirty million years to get it exactly this wrong-footing and this good. You do not build a machine like that in a hurry. You cannot. The whole strategy is to outlast the difficulty rather than to overpower it, and that is the first face of the long game.

The partner it cannot see

Under the con sits the thing the con is built on, and it is the deeper of the two stories, and it is the one I most want you to carry out of this book.

You met the partner in the dark, threaded into the orchid's own root cells: the fungus that finds the dust-grain seed where it falls, opens it, feeds it through years of buried infancy, and never quite lets go even when the grown plant has green leaves of its own. The lady's slipper does not feed itself the way a tomato does. It lives, from its first hours and on through every flowering year, half on sunlight and half on a partner it did not choose, cannot see, has no way to hold onto, and very likely cannot fully repay. I asked you, back at the start, to resist the easy word for that. We call that kind of dependence fragility. I tried to convince you it is better called trust, and I meant the hard kind, the kind with no leverage and no proof and no fallback, where the giving-over is the whole of how you live. The orchid has staked its entire existence on something it cannot account for, and the stake has held, in one form or another, for thirty million years.

That is the second face of the long game, and it is the one that took me longest to understand, standing on my rise. The flower I go back to each spring is not, by itself, the living thing. The living thing is the flower and the few inches of unremarkable soil under it, and the fungus laced through that soil finer than thread, and the slow exchange running between them out of all sight, the fungus building its coils and the orchid eating them and the fungus building them again. I have looked at that flower for most of my life and never once seen the half of it that matters most. None of us has. Across the great northern body of this plant's range, nobody has even named the fungus that feeds it. We owe these orchids their whole existence to a partner we are still only learning to see, in places we have not yet looked, and that not-knowing is not a gap in the wonder. It is the wonder, still going.

Here is what the partnership taught me, slowly, on the rise. I had it backward at the start, the way most of us have it backward. I thought the impressive thing about a living creature was how little it needed, how much it could do alone. The lady's slipper makes a quieter and truer case. Nothing alive is alone. We are all of us fed and held and kept going by partners we did not choose and mostly never see, threaded so far into our own bodies and days that we have stopped noticing them. The orchid is not the strange exception to that rule. It is the plain rule, pulled up where you can finally look at it, wearing its dependence at the root. To kneel in front of one and understand what is happening underground is to be handed, free, a corrected picture of how everything lives.

You cannot move that much time

The third thread is time, and the orchid holds more of it than almost anything you will meet on a walk.

Remember what the visible plant actually is. The pouch nodding on its scape is the youngest, briefest, most expensive part of a single individual that may have been spreading and storing and waiting underground for decades, far longer than the flower, sometimes longer than the forest standing over it. The stem in your hand is the part of a very old life that happens to be showing this week. A lady's slipper can take the better part of a decade to climb from a dust-grain seed to a first bloom, and then go on for a human lifetime or more, drawing down a reserve in its rhizome that, once spent, never refills. The flower is the brief exception. The plant is the long, patient rule.

This is why you cannot dig one up and keep it, and I am not going to relitigate that here either, except to name the reason that runs deepest. It is not only that your shovel cannot lift the fungus, though it cannot. It is that your shovel cannot lift the time. You cannot carry home the decades of accumulated underground growth that built the plant, any more than you can transplant the age of a tree by moving the tree. That is the whole of it, and I will say it plainly one last time: a wild lady's slipper is one of the few things on earth you cannot have by taking it. Take it and you do not get the plant. You get its death, on a delay, and a hole in the ground where it stood. You can truly have one only by leaving it exactly where it is, or by making one, the slow way, feeding a dust seed by hand in a sterile flask the years a fungus would have fed it. Leave it, or make it. Those are the only two ways the wanting has to go.

I felt the wanting myself, years ago, kneeling at a pink one with a trowel in my pack, doing the easy arithmetic of how few minutes it would take and how good it would look at home. What stopped my hand was not a rule. It was the slow understanding that the thing I wanted to carry off was not portable, that what made it itself was fastened to that exact slope and that exact soil and that exact partner in ways I could neither see nor lift, and that the flower in front of me was already, completely, where it belonged. The wanting did not vanish. It turned into something closer to gratitude, and I left the plant where it stood. That is the third face of the long game, and it is the one with a lesson sharpest aimed at us: some things can only be had by being left alone, and the leaving is not a loss. It is the only way to keep them.

A standing agreement among conditions

Pull all the way back now, off the single rise, up to where you can see the whole arrangement at once, because there is one more way of holding it that gathers the others in.

A lady's slipper is not really a thing. It is an agreement. It is what gets produced, and goes on being produced, while a long list of conditions all hold at the same time: the soil chemistry sour enough or sweet enough, the water table at the right depth, the light coming through at the right strength, the disturbance regime, the fire that has to have come through and then stayed away, the fungus present in the ground, the particular bee still flying when the flower opens. Hold all of those at once and the agreement keeps producing the orchid, spring after spring. Let one clause quietly lapse and the orchid stops, not with any drama, often with no visible cause at all. The water table drops a foot. The canopy closes. The fire that kept the prairie open stops coming. Even the air gets cleaner, and somewhere a sour-ground specialist loses the very harshness that was protecting its partner. The plant rarely dies because something dramatic happens to it. It dies because one clause of a long contract goes unmet, and the agreement, lacking that clause, simply stops being kept.

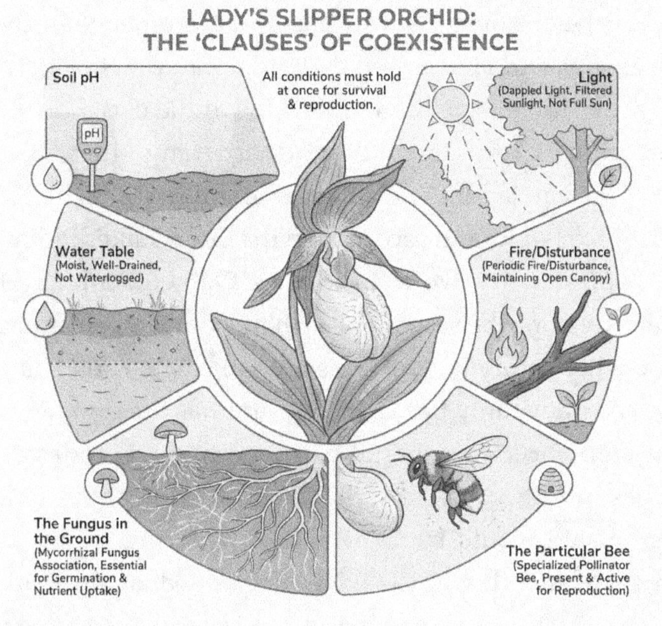

A lady's slipper as a standing agreement among conditions: soil chemistry, water, light, fire, fungus, and pollinator, all holding at once. Let one clause lapse and the flower quietly stops.

This is the frame I find I reach for most, now, standing anywhere these flowers grow, because it tells you what you are actually looking at and what it would actually take to lose it. The pink one on my rise is the standing output of an agreement among that hillside's acid soil, its filtered pine light, its undisturbed litter, the fungus in its duff, and the queen bumblebee who comes up out of the ground each spring still timed to its bloom. Every spring I find it, all of those clauses held for one more year. The flower is the receipt. And the thing about a contract with that many clauses, fitted that precisely to one patch of the world, is that it leaves no room to spare. A narrow door is the most selective thing there is, and the most easily shut. Precision in a changing world is a kind of exposure. That is not a reason for despair. It is a reason for attention, which is the only thing any of this has ever asked of us.

The name it has carried

The flower I keep going back to has a name older than the one the botanists gave it, and I have used that older name all through this book, so let me set it down once more, carefully, at the close.

Across much of this orchid's northern range it is the moccasin flower, a name carried in Anishinaabemowin in forms built from *makizin*, moccasin or shoe, and recorded in the Great Lakes Indian Fish and Wildlife Commission's work and in the Ojibwe People's Dictionary. There is an Anishinaabe story held with this flower, kept and carried by the people whose story it is, and it is not mine to tell and I have not told it. I name the name only because the plant has lived in language, and in relationship, far longer than it has lived in field guides or in books like this one, and because to write a whole book about a flower and leave that unsaid would be its own kind of taking. I am a settler and a learner here, standing at the edge of a knowledge that is not mine, naming what I can rightly name and leaving the rest where it belongs, with its keepers.

That much I can carry to my rise and set down beside the orchid without presuming on anything. The flower has been somebody's relation far longer than it has been my curiosity. Knowing that does not make it less mine to love. It makes the loving more careful, which is the only direction this whole book has ever pushed.

What the long game asks

So here is what the orchid has been teaching me, on that low rise, for most of a life, without ever once saying it out loud.

It has taught me that the impressive thing is staying, not speed, and that the things you care about most are the ones you keep only by leaving them where they stand. None of that arrived as a lesson. It arrived as a flower, opening on the same rise in the same sour soil every May, while I came and went and aged and finally slowed down enough to notice.

And I notice now that I have not, in all these years, done anything to that plant. I have never dug it, never picked it, never moved it, never told anyone exactly where it is. I have given it a wide circle of untrodden ground and stayed at the edge of it. The whole of my relationship with this flower has been to find it, look at it, and leave it alone. That is not nothing. The orchid asked nothing of me but attention and restraint, which are the two halves of love that does no harm, and it has been quietly teaching me both for as long as I have known it.

This is where the book has been going the whole time. Not to a rule you could put on a sign. To a kind of looking. By now, I hope, you cannot walk past one of these without stopping, and when you stop you see the pink, and you think *one-way valve* at the slit down the pouch, and your eye goes down past the flower to the few inches of ordinary soil where the partner is, and you feel, standing there, the press of all that patient time the plant is made of and you are not. The care is just what the wonder turns into, once you understand what you are looking at.

In the pine woods

It is the last warm afternoon of May, and I am back on the rise.

The light is coming through the half-opened oaks in that thin green way it does for only a few weeks a year, and the pine duff is dry and springy under my knees, and there, between the two pleated leaves pressed close to the ground, is the bare scape, and the pouch at the top of it, pink and veined and faintly swollen, nodding in air too still to feel. I found it again. Another spring, every clause of the long agreement held for one more year. I do not touch it. I have never touched it. I crouch a careful arm's length back, on ground already packed by my own past visits so I am crushing nothing new, and I look at the slit down the front of the pouch, and I look at the litter at its base where the fungus is working in the dark, and I let myself feel, the way I always do now, that the half of this plant that matters most is the half I cannot see, and that it was here before me and will be here after, keeping its own slow time.

Somewhere below the rise a bumblebee is working the warm edge of the woods, and the flower is waiting for her, patiently, with the deep patience of a lineage that has been at exactly this for longer than there have been people to kneel down and watch.

I stay crouched a while longer. Then I get up, and brush the pine needles from my knees, and leave it where it stands.

The last warm afternoon of May, and the flower is found again. One more spring, every clause of the long agreement kept.

A Gallery of the Twelve

Here are all twelve native lady's slippers in botanical art: six watercolors painted by Mary Vaux Walcott between 1904 and 1924, now held by the Smithsonian American Art Museum; four hand-colored nineteenth-century plates from *Curtis's Botanical Magazine*; and two watercolor-style renderings made for this book, with the aid of AI, for the species no historical painter left us.

Pink lady's slipper, Cypripedium acaule. *Watercolor by Mary Vaux Walcott, 1904. Smithsonian American Art Museum, CC0.*

Yellow lady's slipper, Cypripedium parviflorum. *Watercolor by Mary Vaux Walcott, 1920. Smithsonian American Art Museum, CC0.*

Kentucky (ivory) lady's slipper, Cypripedium kentuckiense. *Watercolor-style rendering made for this book with AI; no historical watercolor of this species exists.*

Showy lady's slipper, Cypripedium reginae. *Watercolor by Mary Vaux Walcott, 1924. Smithsonian American Art Museum, CC0.*

Small white lady's slipper, Cypripedium candidum. *Hand-colored plate drawn and lithographed by Walter Hood Fitch,* Curtis's Botanical Magazine, *1870. Public domain.*

Ram's-head lady's slipper, Cypripedium arietinum. *Watercolor by Mary Vaux Walcott, 1924. Smithsonian American Art Museum, CC0.*

Mountain lady's slipper, Cypripedium montanum. *Watercolor by Mary Vaux Walcott, 1923, the opening plate of her* North American Wild Flowers. *Smithsonian American Art Museum, CC0.*

California lady's slipper, Cypripedium californicum. *Hand-colored plate drawn by Matilda Smith and lithographed by John Nugent Fitch,* Curtis's Botanical Magazine, *1891. Public domain.*

Clustered lady's slipper, Cypripedium fasciculatum. *Hand-colored plate drawn by Matilda Smith and lithographed by John Nugent Fitch,* Curtis's Botanical Magazine, *1893. Public domain.*

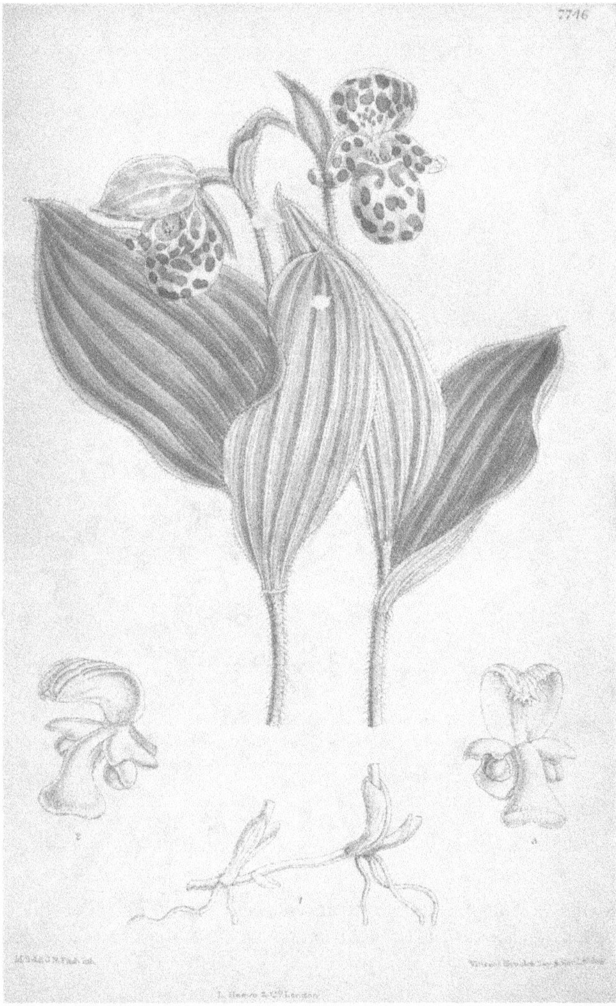

Spotted lady's slipper, Cypripedium guttatum. *Hand-colored plate drawn by Matilda Smith and lithographed by John Nugent Fitch,* Curtis's Botanical Magazine, *1900. Public domain.*

Palomino lady's slipper, Cypripedium yatabeanum. *Watercolor-style rendering made for this book with AI; no historical watercolor of this species exists.*

Sparrow's-egg lady's slipper, Cypripedium passerinum. *Watercolor by Mary Vaux Walcott, 1916. Smithsonian American Art Museum, CC0.*

Acknowledgments

My deepest thanks to my mother, Kim Zyla-Scott, and my wife, Courtney Zyla, without whom none of this would have been written.

This book also stands on the patient work of others: the botanists, mycologists, and field researchers whose published studies are its factual backbone; the native-plant societies and natural-heritage programs that keep these orchids, and their locations, safe; and the photographers and institutions who made their images freely available. The Indigenous Nations and knowledge-holders whose long relationships with these flowers predate every field guide are named, where the published record allowed, in the pages themselves. I am grateful to have learned from the record they and their allies have built, and I hold myself open to correction wherever I have fallen short.

Where to See Them, by Region

A word on how to use this. What follows is a guide to which lady's slippers live where, and the kinds of places to look for them — not a list of plants to go dig up, and not a set of map pins. For a rare orchid the precise location is the most dangerous thing about it, so the right way to find a wild one is through the people who already keep its address responsibly: your state or provincial natural-heritage program, a regional native plant society, or a local botanical garden. Many of them run guided walks in season. And the surest way to see these flowers without risk to them is to visit a managed public preserve or wildflower garden, where they grow on marked trails behind the protection of stewardship.

Bloom runs roughly from late April in the South to July in the far North; most of the eastern species peak in late May and June.

The Northeast and Atlantic Canada. The pink lady's slipper (*Cypripedium acaule*) is the one you are likeliest to meet, on dry, acid ground under pine and oak. In wet, lime-rich fens and swamps, look for the showy lady's slipper (*C. reginae*) and the yellow (*C. parviflorum*); in cold white-cedar swamps, the strange little ram's-head (*C. arietinum*). On the limestone of the Mingan Archipelago in Quebec grows the sparrow's-egg lady's slipper (*C. passerinum*), at the eastern edge of its mostly northern range.

The Southeast and the Appalachians. The pink lady's slipper again on the acid uplands, and the yellow in richer mountain woods. In the West Virginia mountains the pink runs along the heath and dry pine-oak ridges of the Allegheny Plateau, the Dolly Sods country among them; the yellow keeps to the rich coves below; and a few cold, high-elevation limestone fens still shelter the showy lady's slipper at some of its last sites in the state. In the wooded floodplains of the Gulf states and the Ouachitas, the ivory-flowered Kentucky lady's slipper (*C. kentuckiense*) holds the largest pouch of any of them, a rare and local find.

The Upper Midwest and Great Lakes. Perhaps the best region of all for the genus. The yellow and showy lady's slippers in fens and wet woods, the pink on sandy and boreal ground, the ram's-head in cedar swamps and on alvar and dune, and the small white lady's slipper (*C. candidum*) in the last unplowed wet prairies.

The Prairies and Plains. Here the small white lady's slipper is the prize, a plant of intact tallgrass prairie and prairie fen, with the prairie form of the yellow alongside it. The Manitoba Tall Grass Prairie Preserve is its great northern stronghold.

The Mountain West and the Rockies. The mountain lady's slipper (*C. montanum*), white-pouched and dark-petaled, ranges widely from the foothills upward; the yellow reaches here too; and in cool conifer forest, the small, dusky clustered lady's slipper (*C. fasciculatum*) hides so well that the great wildflower books once passed it by.

The Pacific Coast. In the serpentine seeps of northwestern California and southwestern Oregon, often beside the cobra lily, grows the California lady's slipper (*C. californicum*), with its tall raceme of small white pouches. The mountain and clustered lady's slippers share the wider region.

The Far North. Across boreal Canada you may find the spotted lady's slipper (*C. guttatum*), the sparrow's-egg, the boreal yellow, and the pink; the first three reach on into Alaska, the pink does not. And on the treeless Aleutians and the Alaska Peninsula, almost nowhere else in the United States, grows the palomino-flecked C. *yatabeanum*, the hardest of them all to reach.

Public gardens and preserves. The kindest way to meet these orchids is where someone is already looking after them. Botanical gardens with native collections, fens and bogs with boardwalk trails, and managed preserves let you see lady's slippers up close without setting foot on their roots. A few with a long history of it: the Wild Gardens of Acadia in Maine, and, in Manitoba, the Agassiz Trail at the Tall Grass Prairie Preserve and the boardwalk at the Brokenhead Wetland Ecological Reserve. Your regional native plant society will know the nearest.

Where to start. For any wild population, begin with **NatureServe Explorer** (explorer.natureserve.org) and your state or provincial heritage program, or a local native plant society — and keep what you find the way they do, off the open internet.

How to Support Conservation

If the book has done its work, you want to help. Here is what actually helps these orchids, roughly in order of how much good it does.

Leave wild plants where they are, and buy only nursery-propagated stock. This is the first and largest thing. A dug wild lady's slipper almost always dies, and its loss is permanent. If you want to grow one, buy only plants raised from seed in a laboratory flask, never wild-collected stock. Ask the seller directly how the plant was propagated; a flowering-size plant priced too cheaply to have been lab-raised almost certainly was not. A handful of specialist nurseries, Spangle Creek Labs among them, raise lady's slippers honestly this way.

Report poaching. If you see someone digging or selling wild orchids, report it. In the United States the Fish and Wildlife Service tip line is **1-844-FWS-TIPS** (1-844-397-8477), and you can remain anonymous; a state conservation officer is often faster, being closer to the scene. Collecting from public land, and trafficking across state or national lines, is illegal, and the cases on record show the laws are enforced.

Become another pair of eyes. Careful amateur observation is genuinely useful to the science, and most of these orchids are under-watched. A few ways in:

- Record what you find on **iNaturalist**, which automatically blurs the location of rare orchids so your photograph helps without giving the plant away. If you would like your photos usable by researchers and by books like this one, set their license to CC0 or CC-BY in your account settings.
- Log **phenology**, the dates a plant buds, opens, and sets seed, through programs like the USA National Phenology Network's *Nature's Notebook*. Returning to the same plant week by week across a season produces exactly the data that is thinnest for *Cypripedium* and most needed as the climate shifts.
- Join a rare-plant monitoring program. The Native Plant Trust's volunteers cover New England; Wisconsin, California, and many other states and provinces run their own. The Smithsonian's North American Orchid Conservation Center coordinates a continent-wide pollinator survey, and the pollinators of roughly half our native orchids have still never been seen.

Support the organizations doing the work. Membership, a donation, or a day of volunteering all help. Among those active on native lady's slippers: the **North American Orchid Conservation Center** (a Smithsonian-led network), the **Center for Plant Conservation**, the **Native Plant Trust**, the **California Native Plant Society**, **Native Orchid Conservation Inc.** in Manitoba, the **Native Orchid Conference**, and **United Plant Savers**, which tracks at-risk native plants. Your state or provincial natural-heritage program is the institution that actually keeps the records, and most welcome both reports and support.

Protect the ground itself. In the end a lady's slipper is an agreement among conditions, and the surest way to keep one is to keep its habitat whole. Support the land trusts, preserves, and heritage programs that protect fens, bogs, prairies, and old acid woods, and the prescribed fire and shifting water tables that some of these orchids cannot live without. Saving the place saves the flower, the fungus, and the bee all at once.

Glossary

A quick reference to the botanical and ecological terms this book uses, defined as the book uses them.

Acidophile. A plant that needs sour, acidic ground, the lower the pH the better. The pink lady's slipper is the genus's signature acidophile, at home in soil too acid (around pH 4 to 4.5) for most other plants to tolerate. (See Calciphile.)

Anther. The part of a flower where its own pollen is made and held. An escaping bee passes the lady's slipper's anther only after she has passed the stigma, so she leaves another flower's pollen behind before she picks up a fresh load. (See Column, Stigma.)

Asymbiotic germination. Sprouting an orchid seed in the laboratory on a sterile nutrient gel, with no fungus present, so the gel feeds the seedling the way a fungus would in the wild. The breakthrough that made raising lady's slippers from seed possible (Lewis Knudson, 1922; adapted for lady's slippers by Curtis in 1943). (See Green-pod technique, Deflasking.)

Autogamy. Self-pollination: a flower fertilizing itself with its own pollen, with no bee or other go-between. Most lady's slippers work hard to avoid it; the sparrow's-egg lady's slipper (*Cypripedium passerinum*) is the genus's routine self-pollinator. (See Geitonogamy.)

Bilateral symmetry. The build of a flower that has a left and a right mirroring each other, like a face, so there is only one way to meet it head-on. Orchids share it with snapdragons and peas, and it is part of why a lady's slipper seems to look back at you.

Calciphile. A lime-lover: a plant that needs sweet, calcium-rich, alkaline ground and will not grow without it. The small white lady's slipper (*Cypripedium candidum*) and the showy lady's slipper (*C. reginae*) are calciphiles, the mirror image of the acid-loving pink. (See Acidophile.)

Column (gynostemium). The structure at the center of every orchid, where the male and female organs kept separate in most flowers are fused into one. In a lady's slipper the bee must squeeze past the column to escape, which is how the flower gets pollinated. (See Anther, Stigma, Staminode.)

Deceptive pollination. The orchid's central trick: advertising a meal, with color and scent and the shape of a generous flower, that the plant never intends to serve. The bee is drawn in, finds nothing, and leaves having worked for free. (See Generalized food deception.)

Deflasking. Moving a lab-raised orchid seedling out of its sterile glass flask and into ordinary soil. It is the most dangerous step in propagation, the door between the glass and the ground, where most young plants are lost. (See Asymbiotic germination.)

Dust seed. The orchid's seed: barely more than a speck, about a millimeter long, carrying almost none of the packed food a bean or an acorn gives its sprout. A lady's slipper makes them by the thousand and casts them on the wind, betting that a few will land where the right fungus waits. (See Endosperm, Protocorm.)

Ectomycorrhizal. Describing a fungus that sheathes the root tips of forest trees such as conifers, trading minerals for the trees' sugar. A few lady's slippers tap this tree-and-fungus network; the clustered lady's slipper (*Cypripedium fasciculatum*) draws much of its carbon from it. (See Mycorrhiza, Partial mycoheterotrophy.)

Endosperm. The food supply built into an ordinary seed (the white of a corn kernel, the meat of a coconut) to feed the sprout until it can feed itself. Orchid seeds have none, which is why they cannot grow without a fungus. (See Dust seed, Mycoheterotrophy.)

Genet and ramet. Two words for what looks like one plant. A ramet is a single visible stem; a genet is the whole genetic individual, the original plant plus every stem it has sent up by creeping underground, all one continuous living thing. A clump of a dozen lady's slipper stems may be a single, very old genet. (See Rhizome, Vegetative dormancy.)

Geitonogamy. A subtler self-pollination, in which an insect carries a flower's pollen to another bloom on the same plant. It counts as inbreeding, and the lady's slipper's rewardless strategy is partly built to prevent it: a cheated bee leaves in a hurry and carries the pollen to a different plant. (See Autogamy.)

Generalized food deception. The particular con most lady's slippers run: imitating the general look and scent of a rewarding flower without copying any single real one, so an inexperienced bee expects a meal that is not there. (See Deceptive pollination.)

Geoprivacy. On nature-sharing platforms such as iNaturalist, the setting that hides a sensitive plant's exact location, showing it only inside a wide blurred circle. For a rare orchid, whose precise address is the most dangerous fact about it, turning it on is a basic kindness. (See Obscuring.)

Glandular. Of a hair or surface: tipped with a tiny gland that weeps a sticky or irritating fluid. The showy lady's slipper's stem and leaves are covered in glandular hairs that can raise a rash on bare skin, a defense disguised as softness.

Green-pod technique. Harvesting an orchid's seed pod while it is still green and unripe, weeks after pollination, and sowing the immature seed at once. The not-yet-hardened seed sprouts far more readily than ripe seed, and the method underlies most lady's slipper propagation today. (See Asymbiotic germination.)

Incipient species. Populations caught in the act of splitting into separate species, partway through and still trading genes across the widening gap. Several varieties of the yellow lady's slipper look like incipient species, a flower caught mid-sentence. (See Reticulate evolution.)

Kettle trap. The shared design of every lady's slipper pouch: a chamber an insect enters easily but cannot leave the way it came, escaping only by one forced route past the flower's reproductive organs. Named for the old one-way fish trap built on the same principle. (See Labellum.)

Labellum. The botanists' name for the pouch, or slipper: a single petal swelled into a hollow chamber, the part that gives the lady's slipper all its common names and around which the whole flower is arranged. This book mostly just calls it the pouch.

Mixotroph. A mixed feeder: a plant that lives partly on sunlight, through its own green leaves, and partly by feeding on a fungus underground. Adult lady's slippers stay mixotrophs for life, never fully weaning off their fungal partner. (See Mycoheterotrophy, Partial mycoheterotrophy.)

Mycoheterotrophy. Living by eating a fungus. In its first years underground, before it has a single green leaf, a lady's slipper is wholly mycoheterotrophic, fed entirely by the fungus laced into it. (See Protocorm, Mixotroph.)

Mycorrhiza. Literally "fungus-root": the close partnership between a plant's roots and a fungus in the soil, grown so far together that each feeds the other what it cannot get alone. In a lady's slipper the fungus lives inside the root cells, and the plant cannot survive without it. The central relationship of this book. (See Peloton.)

Obscuring. The long practice, among the programs that track rare plants, of recording a sensitive location only in general terms (a county, a watershed, a coarse dot) rather than a findable point. (See Geoprivacy.)

Pandurate. Fiddle-shaped: of a petal, pinched in near the middle like the waist of a violin. The lateral petals of *Cypripedium yatabeanum* are pandurate, the mark that tells it under a lens from the plain narrow petals of its near-twin, the spotted lady's slipper (*C. guttatum*).

Paraphyletic. Of a named group: not a clean, closed circle, because it leaves out some of its own descendants or takes in an outsider. The lady's slippers prove paraphyletic in places, part of why naming the species is so hard. (See Reticulate evolution.)

Partial mycoheterotrophy. A halfway state in which a green, leafy plant still pulls a large share of its carbon up through fungi instead of making all of it in its leaves. The clustered lady's slipper takes nearly half its carbon this way, the mixotroph idea carried to its extreme. (See Mixotroph, Ectomycorrhizal.)

Peloton. A coil of fungus packed inside a single orchid root cell, wound like a skein of thread. The orchid digests these coils for food and the fungus grows them back, a slow cycle of building and eating that keeps the whole partnership running. (See Mycorrhiza.)

Phenology. The timing of a plant's year: when it breaks ground, buds, opens, and sets seed. Following the phenology of one lady's slipper, week by week across a season, is the most useful single thing an amateur can add to the science.

Protocorm. The tiny underground swelling a germinated orchid seed becomes on its way to a first leaf, fed entirely by its fungus, sometimes for years, before any green shows above the soil. (See Dust seed, Mycoheterotrophy.)

Reticulate evolution. Evolution that runs not as a tidy branching tree but as a net, with separate lineages reaching back across the gaps to fuse and braid rather than only dividing. The lady's slippers are a clear case, which is why their family tree resists a clean drawing. (See Paraphyletic, Incipient species.)

Rhizome. The thick horizontal stem a lady's slipper grows just below the surface, creeping along year by year. It is the plant's pantry, anchor, and long memory, the store it lives on, and a chief reason a dug-up plant so seldom survives. (See Genet and ramet, Vegetative dormancy.)

Scape. A flowering stalk that rises straight from the ground bearing no leaves, the leaves staying gathered at the base. The pink lady's slipper's pouch nods atop a bare scape; its scientific name, *acaule*, means "stemless."

Sepal (dorsal sepal and synsepal). The outer parts of a flower, wrapping it in bud. A lady's slipper carries a showy dorsal sepal like a hood above the pouch and, in most species, two lower sepals fused into one synsepal behind it. The ram's-head lady's slipper (*Cypripedium arietinum*) is the only native that leaves its lower sepals separate.

Serpentine. Rock carried up from deep in the earth's mantle, and the strange soil it weathers into: heavy in magnesium and metals, poor in calcium, and close to poison for most plants. The California lady's slipper (*Cypripedium californicum*) is fixed to serpentine seeps that almost nothing else will grow in.

Staminode. A sterile, modified stamen that makes no pollen. In a lady's slipper it is the small shield, often colored or spotted, that roofs the mouth of the pouch and helps steer the bee past the working parts. (See Column.)

Stigma. The receptive surface that catches and holds incoming pollen. In a lady's slipper the escaping bee meets the stigma first and the anther second, so she delivers another flower's pollen before collecting this one's, the arrangement that forces cross-pollination. (See Anther, Column.)

Testa. The hard protective coat around a seed. The orchid's dust seed wears only a thin one, part of why it weighs so little and carries so little food. (See Dust seed.)

Vegetative dormancy. A living lady's slipper's habit of staying wholly underground for a season or more, sending up no stem and no leaf, then reappearing as if it had never left. It rides out those years on its rhizome and its fungus, which is why a plant that disappears has not necessarily died. (See Rhizome.)

Selected Sources and Further Reading

This book carries no footnotes. Where a fact comes from a particular person or study, I have named them in the text, in the belief that a name read in a sentence does more for a curious reader than a number sending them to the back of the book. What follows is not the full apparatus behind these pages, which runs to well over a thousand entries, but a short, opinionated guide to the sources I leaned on most, and the places to go if you want to know more.

For looking up any species. The authoritative taxonomic treatment is Charles J. Sheviak's account of *Cypripedium* in the *Flora of North America*, volume 26 (Oxford University Press, 2002), free online at efloras.org. The standard monograph for the genus worldwide is Phillip Cribb's *The Genus Cypripedium* (Royal Botanic Gardens, Kew, and Timber Press, 1997). For range, status, and habitat, NatureServe Explorer (explorer.natureserve.org) and the USDA PLANTS Database are the working tools; the USDA Forest Service's "Celebrating Wildflowers" pages and Go Botany, from the Native Plant Trust, are friendlier first stops.

On how they live. The science moves quickly, and the best single doorway to it is the North American Orchid Conservation Center, a

Smithsonian-led network whose researchers, Melissa McCormick and Dennis Whigham among them, have done much of the modern work on the orchids' fungal partners. The genus's tangled family tree was most recently read by Lagou and colleagues in a 2024 phylogenomic study; the long-term fate of marked plants by Marc Kéry and Katharine Gregg; the founding craft of raising orchids from seed by Lewis Knudson (1922) and, for the slippers, John Curtis (1943). Readers who want the primary literature will find it cited from those starting points.

The voices in these pages. Many of the older observations I quote are public domain and a pleasure to read whole: Susan Fenimore Cooper's *Rural Hours* (1850); the journals of Henry David Thoreau; John Burroughs's *Wake-Robin* (1871) and *Riverby* (1894); Charles Darwin's *Fertilisation of Orchids* (1862, revised 1877); John Muir's *Our National Parks* (1901); Neltje Blanchan's *Nature's Garden* (1900); Catharine Parr Traill's *Canadian Wild Flowers* (1868); John Richardson's botanical appendix to Franklin's *Narrative of a Journey to the Shores of the Polar Sea* (1823); and Mary Vaux Walcott's *North American Wild Flowers* (Smithsonian, 1925), where several of these orchids were first painted for a wide audience.

On Indigenous plants and knowledge. I have tried to name Nations and published sources in the text rather than gesture at "Native peoples," and the sources I relied on are worth seeking out directly: James Mooney's nineteenth-century Cherokee ethnography; Frank Speck's work with the Wabanaki; Frances Densmore's *Uses of Plants by the Chippewa Indians* (1928); Daniel Moerman's *Native American Ethnobotany* (Timber Press, 1998); the Ojibwe People's Dictionary at the University of Minnesota; the Great Lakes Indian Fish and Wildlife Commission's *Indinawemaaganidog / All My Relatives* (2007); and Mary Siisip Geniusz's *Plants Have So Much to Give Us* (University of Minnesota Press, 2015). On how a settler should handle this knowledge at all, I am indebted to Robin Wall Kimmerer's *Braiding Sweetgrass* (Milkweed, 2013) and to the frameworks of Local Contexts, the CARE

Principles for Indigenous Data Governance, and the OCAP principles in Canada.

If you liked this book. For more in this vein, science told as wonder, I would send you to Kimmerer's *Braiding Sweetgrass* again, to Zoë Schlanger's *The Light Eaters* (2024), and, for the human obsession that orchids provoke, to Susan Orlean's *The Orchid Thief* (1998).

Illustration and Photo Credits

This book carries no footnotes, and most of its sources are named in the text. Its images are credited here instead. Several of the photographs, all of the diagrams and maps, and two of the twelve plates in the gallery were made with the aid of artificial intelligence; the rest are historical art, public-domain documents, openly licensed photographs, or the author's own pictures. Where a license requires a printed credit, it is given in full below. Public-domain and CC0 items carry no legal requirement of attribution, and are credited here as a courtesy and a record.

A note on the AI-generated images

Every photograph, diagram, and map in this book that is not credited to a named source in the sections below was generated with the aid of artificial intelligence, using Google's Gemini 3 Pro Image model, and was then checked for botanical accuracy by the author against the scientific sources behind the text. The twelve species portraits, the diagnostic close-up details, and two photographs of orchid behavior (a bumblebee and a crab spider) are licensed photographs by the photographers credited below, not AI images. What remains AI-made is the habitat and atmosphere scenes, all of the diagrams, maps, and infographics, and two of the twelve plates in the gallery: the Kentucky lady's slipper and the palomino lady's slipper, the two species for which no historical painting exists. The use of AI in making these images is disclosed at the point of sale and in the book's copyright registration.

Botanical art (public domain)

- Six watercolors by Mary Vaux Walcott, painted between 1904 and 1924 and reproduced in the gallery: the pink, yellow, showy, ram's-head, mountain, and sparrow's-egg lady's slippers. Smithsonian American Art Museum, released under CC0 through Smithsonian Open Access.
- Four hand-colored plates from *Curtis's Botanical Magazine*, public domain: the small white lady's slipper (plate 5855, 1870, drawn and lithographed by Walter Hood Fitch); and the California lady's slipper (plate 7188, 1891), the clustered lady's slipper (plate 7275, 1893), and the spotted lady's slipper (plate 7746, 1900), each drawn by Matilda Smith and lithographed by John Nugent Fitch.

Historical figures, portraits, and documents (public domain)

- Charles Darwin's diagram of *Cypripedium* pollination, from *The Various Contrivances by which Orchids Are Fertilised by Insects*, second edition (1877).
- Albert Bernhard Frank (1839–1900), who coined the word *mycorrhiza*; portrait, 1903.
- Ryōkichi Yatabe (1851–1899), the botanist for whom the palomino lady's slipper is named; photograph by Ogawa Kazumasa, 1899.
- A female *Lasioglossum* sweat bee; USGS Native Bee Inventory and Monitoring Lab (Sam Droege). Public domain, as a work of the United States government.
- Beaded hide moccasins attributed to an Anishinaabe (Ojibwe) maker, about 1890 to 1910. Minneapolis Institute of Art, The Frances M. Norbeck Fund (2000.138.4a,b).
- The opening page of Elaine Goodale, "Moccasin Flower," in *In Berkshire with the Wild Flowers* (1879).

- The *Minneapolis Journal* editorial on the state flower, 22 February 1902. Library of Congress, Chronicling America.
- Constantine Rafinesque, *Medical Flora,* volume 1 (1828): the "American Valerian" plate.
- A yellow lady's slipper from Charles Millspaugh, *American Medicinal Plants* (1887), plate 170.

Photographs under Creative Commons

CC0 (no rights reserved; credited as a courtesy):

- The potted *Paphiopedilum* hybrid in Chapter 2. Photograph by Agnes Trekker, via iNaturalist.

CC BY 4.0 (creativecommons.org/licenses/by/4.0/):

- Scanning-electron micrograph of an orchid dust seed (*Bletilla striata*). Miura and colleagues, *Plants* (2019), CC BY 4.0.
- Fungal pelotons in an orchid root cell (*Pleione*). Qin and colleagues, *Journal of Fungi* (2021), CC BY 4.0.
- *Cypripedium guttatum* seedlings in asymbiotic culture. Park and colleagues, *Plants* (2023), CC BY 4.0.
- A diapriid wasp (*Aneuropria foersteri*), representative of the clustered lady's slipper's pollinator family. Photograph by Jonas V., via iNaturalist, CC BY 4.0.
- *Cypripedium calceolus,* a wild plant in Cumbria, England. Photograph by Nigel Voaden, via iNaturalist, CC BY 4.0.

Species portrait photographs

The photograph that opens each species' account. Public-domain pictures are credited here as a courtesy; the rest carry the attribution their licenses require.

Public domain:

- Pink lady's slipper (*Cypripedium acaule*), Chapter 7. Photograph by Dr. Thomas G. Barnes, U.S. Fish and Wildlife Service.
- Yellow lady's slipper (*Cypripedium parviflorum*), Chapter 8. Photograph by "Hardyplants," via Wikimedia Commons.
- Small white lady's slipper (*Cypripedium candidum*), Chapter 10. Photograph by Mason Brock, via Wikimedia Commons.

- Clustered lady's slipper (*Cypripedium fasciculatum*), Chapter 11. Photograph by the USDA Forest Service, Northern Region (Flathead National Forest).

CC BY (attribution required):

- Showy lady's slipper (*Cypripedium reginae*), Chapter 9. Photograph by NC Orchid, CC BY 2.0, via Flickr.
- Ram's-head lady's slipper (*Cypripedium arietinum*), Chapter 10. Photograph by James Ellison (magnolia1000), CC BY 2.0, via Flickr.
- California lady's slipper (*Cypripedium californicum*), Chapter 11. Photograph by NC Orchid, CC BY 2.0, via Flickr.
- Mountain lady's slipper (*Cypripedium montanum*), Chapter 11. Photograph by Alex Abair, CC BY 4.0, via iNaturalist.
- Palomino lady's slipper (*Cypripedium yatabeanum*), Chapter 12. Photograph by Motohiro Sunouchi, taken in Hokkaidō, Japan, CC BY 2.0, via Flickr.
- Sparrow's-egg lady's slipper (*Cypripedium passerinum*), Chapter 12. Photograph by Dustin Snider, CC BY, via iNaturalist.

CC BY-SA (attribution required, with the ShareAlike condition):

- Kentucky lady's slipper (*Cypripedium kentuckiense*), Chapter 8. Photograph by Eric Hunt, CC BY-SA 4.0, via Wikimedia Commons.
- Spotted lady's slipper (*Cypripedium guttatum*), Chapter 12. Photograph by "Orchi," CC BY-SA 3.0, via Wikimedia Commons.

Diagnostic and behavior photographs

Close-ups of a single feature, and two moments of orchid behavior.

CC BY (attribution required):

- Pink lady's slipper pouch slit (Chapter 7). Photograph by Michael J. Papay, CC BY, via iNaturalist.
- Pink lady's slipper basal leaves and scape (Chapter 1). Photograph by "crystella," CC BY, via iNaturalist.
- Showy lady's slipper staminode (Chapter 9). Photograph by Motohiro Sunouchi, CC BY 2.0, via Flickr.
- A crab spider in a sparrow's-egg lady's slipper pouch (Chapter 10). Photograph by James Telford, CC BY, via iNaturalist.

CC BY-SA (attribution required, with the ShareAlike condition):

- Yellow lady's slipper twisted lateral petals (Chapter 8). Photograph by "Orchi," CC BY-SA 3.0, via Wikimedia Commons.
- A bumblebee entering a pink lady's slipper (Chapter 7). Photograph by Aaron Carlson, CC BY-SA 2.0, via Flickr.

Photographs by the author and his mother

- The author's own *Cypripedium* 'Philipp Dark,' a nursery-bred Kentucky lady's slipper hybrid, in Chapter 14. Photograph by the author; all rights reserved.
- A cultivated yellow lady's slipper (*Cypripedium parviflorum*) in the garden of the author's mother, in Chapter 8. Photograph by Kim Zyla-Scott; all rights reserved.

About the Author

Joel Zyla is not a botanist. He is a lifelong lover of plants, orchids most of all, who was first led into the woods as a boy by his grandfather and taught to look. He lives in West Virginia, where several of these orchids grow wild (the pink one on the sour ridges, the showy in a few cold limestone fens), and a Coast Guard career that took him from the Great Lakes to the northern California coast introduced him to more of them in the wild. When he is not looking for lady's slippers he is often tending fungi instead, on inoculated logs and a wood-chip bed behind the house. He would rather kneel down to photograph a lady's slipper than do almost anything else.

Index

Page numbers refer to the print edition.

www.ingramcontent.com/pod-product-compliance
Lightning Source LLC
LaVergne TN
LVHW022154200926
848481LV00006B/1737
9798996547814